T0220847

Marine Environmental Characterization

Marine Environmental Characterization
C. Reid Nichols and Kaustubha Raghukumar

ISBN: 978-3-031-01362-1 paperback
ISBN: 978-3-031-02490-0 ebook
ISBN: 978-3-031-00318-9 hardcover

DOI 10.1007/978-3-031-02490-0

A Publication in the Springer series
SYNTHESIS LECTURES ON OCEAN SYSTEMS ENGINEERING

Lecture #2
Series Editor: Nikolaos I. Xiros, *University of New Orleans*
Series ISSN
Synthesis Lectures on Ocean Systems Engineering
ISSN pending.

Cover Artwork:
Gulf AUV Network and Data Archiving Long-Term Storage Facility (GANDALF) depicted Gulf Stream System dynamics from its upstream extension known as the Loop Current to meanders and eddies that formed near Cape Hatteras on April 29, 2020. GANDALF is a Gulf of Mexico Coastal Ocean Observing System (GCOOS) capability that provides environmental intelligence to users such as Autonomous Underwater Vehicle (AUV) pilots. This dynamic base map includes information from nautical charts, in situ sensors, weather radar, satellite imagery, and numerical models. (Courtesy of GCOOS, see https://gandalf.gcoos.org/portal).

Synthesis Lectures on Ocean Systems Engineering

Editor
Nikolaos I. Xiros, ***University of New Orleans***

The Ocean Systems Engineering Series publishes state-of-the-art research and applications oriented short books in the related and interdependent areas of design, construction, maintenance and operation of marine vessels and structures as well as ocean and oceanic engineering. The series contains monographs and textbooks focusing on all different theoretical and applied aspects of naval architecture, marine engineering, ship building and shipping as well as sub-fields of ocean engineering and oceanographic instrumentation research.

Marine Environmental Characterization
C. Reid Nichols and Kaustubha Raghukumar
2020

Feedback Linearization of Dynamical Systems with Modulated States for Harnessing Water Wave Power
Nikolaos I. Xiros
2020

Marine Environmental Characterization

C. Reid Nichols
Marine Information Resources Corp., Ellicott City, MD

Kaustubha Raghukumar
Integral Consulting Inc., Santa Cruz, CA

SYNTHESIS LECTURES ON OCEAN SYSTEMS ENGINEERING #2

ABSTRACT

The use of environmental data to support science, technology, and marine operations has evolved dramatically owing to long-term ocean observatories, unmanned platforms, satellite and coastal remote sensing, data assimilative numerical models, and high-speed communications. Actionable environmental information is regularly produced and communicated from quality-controlled measurements and skillful forecasts. The characterization of complex oceanographic processes is more difficult compared to inland features because of the difficulty in obtaining observations from often remote and hazardous locations. Regardless, coastal and ocean engineering projects and operations require the collection and analysis of meteorological and oceanographic data to fill information gaps and the running of numerical models to characterize regions of interest. Data analytics are also essential to integrate disparate marine data from national archives, in situ sensors, imagery, and numerical models to meet project requirements. Holistic marine environmental characterization is essential for data-driven decision making across the science and engineering lifecycle (e.g., research, production, operations, end-of-life).

Many marine science and technology projects require the employment of an array of instruments and models to characterize spatially and temporally variable processes that may impact operations. Since certain environmental conditions will contribute to structural damage or operational disturbances, they are described using statistical parameters that have been standardized for engineering purposes. The statistical description should describe extreme conditions as well as long- and short-term variability. These data may also be used to verify and validate models and simulations. Environmental characterization covers the region where engineering projects or maritime operations take place. For vessels that operate across a variety of seaways, marine databases and models are essential to describe environmental conditions. Data, which are used for design and operations, must cover a sufficiently long time period to describe seasonal to sub-seasonal variations, multi-year, decadal, multi-decadal, and even climatological factors such as sea level rise, coastal winds, waves, and global ocean temperatures. Combined data types are essential for the computation of environmental loads for the region of interest. Typical factors include winds, waves, currents, and tides. Some regions may require consideration of biofouling, earthquakes, ice, salinity, soil conditions, temperature, tsunami, and visibility. Observations are also used for numerical forecasts, but errors may exist due to inexact physical assumptions and/or inaccurate initial data, which can cause errors to grow to unacceptable levels with increased forecasting times. Overall, marine environmental characterization tools, from observational data to numerical modeling, are critical to today's science, engineering, and marine operational disciplines.

KEYWORDS

marine environment, oceanography, numerical models, in situ data, imagery, databases, environmental factors, data analytics, reliability

Dedicated to
Gerald S. Janowitz (1943–2017)
and
Paul A. Wittmann (1956–2019)

Contents

Preface

Marine Environmental Characterization is an introductory geoscience book focusing on physical oceanographic processes. Science issues that are addressed include sea level rise, coastal erosion and hazards, and pollution. Not addressed in this short book are geochemical aspects that relate to corrosion, ocean acidification, and sediment transport, and biological aspects linked to invasive species, marine mammal protection, and overfishing.

Oceanography is approached as a compound discipline integrating geological, physical, chemical, and biological characteristics of the ocean environment. Descriptions of waves, tides, currents, and land-sea-breeze circulation, which are known to even the most junior engineering student, are described. Insights by casual observers such as how breaking waves off the shoreline generate longshore currents and release their energy as surf are introduced in this short textbook and how these factors might influence decision makers. The way these processes contribute to erosion and accretion and maritime mishaps such as capsizing are highlighted. For this reason, this general reference is particularly useful in helping engineering students to understand connections between the geosciences and environmental impacts on engineering projects.

This book provides information on data resources such as satellite altimeter data that can be used for environmental characterization. Recent literature has been cited that provides for further reading and insights on environmental impacts ranging from seakeeping to recurring floods. Data resources provided in this reference book and examples are useful in helping teachers to develop case studies that can help students translate basic knowledge into practice through real-world scenarios. The reader is apprised of the importance of big data, which includes historical information, in situ data, remote sensing, and numerical modeling.

Since the 1872–1876 *Challenger* expedition, scientists have been grappling with big data and the ongoing discoveries and innovations of multidisciplinary scientists have greatly advanced the marine science and technology industry. *Marine Environmental Characterization* emphasizes that big data and Geographic Information Systems (GIS) have revolutionized how scientists can acquire and leverage spatial information. GIS usage has rapidly expanded and is a primary way that environmental intelligence can be developed from disparate data sources.

Marine Environmental Characterization explains geoscience concepts in an engineering context where phenomena are described from the open ocean to the nearshore. State-of-the-art platforms and important monitoring networks that are used to collect meteorological and oceanographic data are introduced along with the need to adopt data standards and implement data quality assurance measures. The book also contains a complementary glossary that describes phenomena of importance to sustainable engineering and coastal resilience.

Gary Zarillo, Ph.D.
Professor, Ocean Engineering and Marine Sciences
Florida Institute of Technology
Melbourne, Florida
April 2020

Acknowledgments

Writing a general reference book on the topic of marine environmental characterization would not have been possible without a review by Dr. Leonard J. Pietrafesa, Professor Emeritus from North Carolina State University, who provided valuable feedback, especially on phenomena common to the South Atlantic Bight. Edits and comments from Dr. Craig Jones, Director of the Marine Sciences and Engineering practice for Integral Consulting in Santa Cruz, CA, greatly improved the final product.

C. Reid Nichols and Kaustubha Raghukumar
May 2020

CHAPTER 1

Introduction

Careful and early attention to environmental issues is essential to all people in the marine science and technology industry. This is especially true since the cycles occurring in nature are complex and not fully understood (e.g., earthquakes, El Niño, and hurricanes). Environmental characterization is a discipline of study that supports the full range of maritime projects from Research, Development, Test, and Evaluation (RDT&E) to operations involving activities such as construction, offshore oil and gas production, shipping, and surveillance. Characterization includes basic and advanced research on the measurement and description of air, land, and water processes and their relationships to process variables or characteristics, baseline environmental conditions, deviations from the baseline, and projections into the future. Environmental characteristics provide critical information to deduce generalized trends in the uniqueness or regularity of environmental qualities for the region of interest.

Environmental processes vary in space and time. In general, spatial variability ranges from millimeters to meters to kilometers to thousands of kilometers while temporal variability ranges from seconds to minutes to hours to days to weeks to seasons to years to decades and onto millennia. Spatial and temporal variability may be attributed to a combination of environmental factors that impact anything from ecology to engineering projects to coastal communities. Figure 1.1 illustrates how this variability occurs across a continuum as a bubble type graphic, where a range of bubble sizes is used to help visualize the variance in environmental factors. The challenge is to collect and synthesize data to recognize how key data elements, parameters, and phenomena are changing in our region of interest. In its most basic form, environmental characterization provides the scientist, engineer, or decision maker with relevant facts and figures that describe the natural features of a particular coastal or marine area of interest. For the marine science and technology industry, environmental conditions of concern are usually those contributing to structural damage, operation disturbances or navigation failures. Figure 1.2 depicts an aerographer mate checking wind speed and direction aboard the USS Theodore Roosevelt on December 12, 2017.

Over the years, many applied scientists, engineers, and mariners have collected and processed large volumes of disparate data to define abiotic and biotic factors common to their region of interest. Numerous authors Bishop [1984], Dickey and Bidigare [2005], Emery and Thomson [2014], Glover et al. [2011], Lein [2012], Pugh [1987] have discussed methods related to the characterization of meteorological and oceanographic data. Information on important phenomena is often described from a combination of historical information, imagery, in situ data,

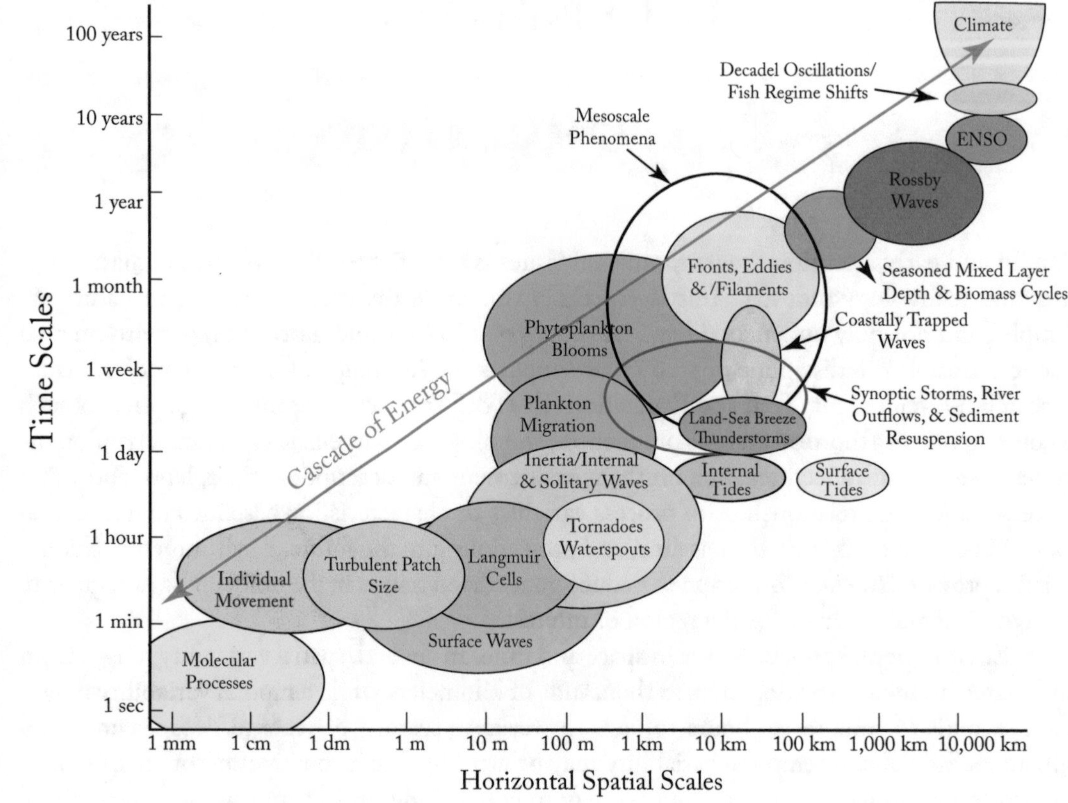

Figure 1.1: Temporal and spatial scales for a variety of ocean processes (adapted from Dickey and Bidigare [2005]).

and numerical model output. Selected phenomena that may be considered as core variables are listed in Table 1.1 and others are provided in the glossary. Example working standards for meteorological and oceanographic variables are listed in Table 1.2. Dimensions and metric units for commonly used parameters are provided in Table 1.3. Government-sponsored programs generally require a data management plan to describe how data are managed and shared throughout the project (e.g., file formats, metadata standards, and conventions).

Many scientists and engineers use various forms of removable memory to store their data and are increasingly moving to cloud storage options to archive data on the internet. The archival and awareness of these data is important because the collections are expensive to obtain and maintain and have value for others. One success in archiving data has been development of databases such as the International Comprehensive Ocean-Atmosphere Data Set (ICOADS), which consist of global ocean marine meteorological and surface ocean measurements from moored and drifting buoys, coastal stations, and ocean platforms and visual observations from

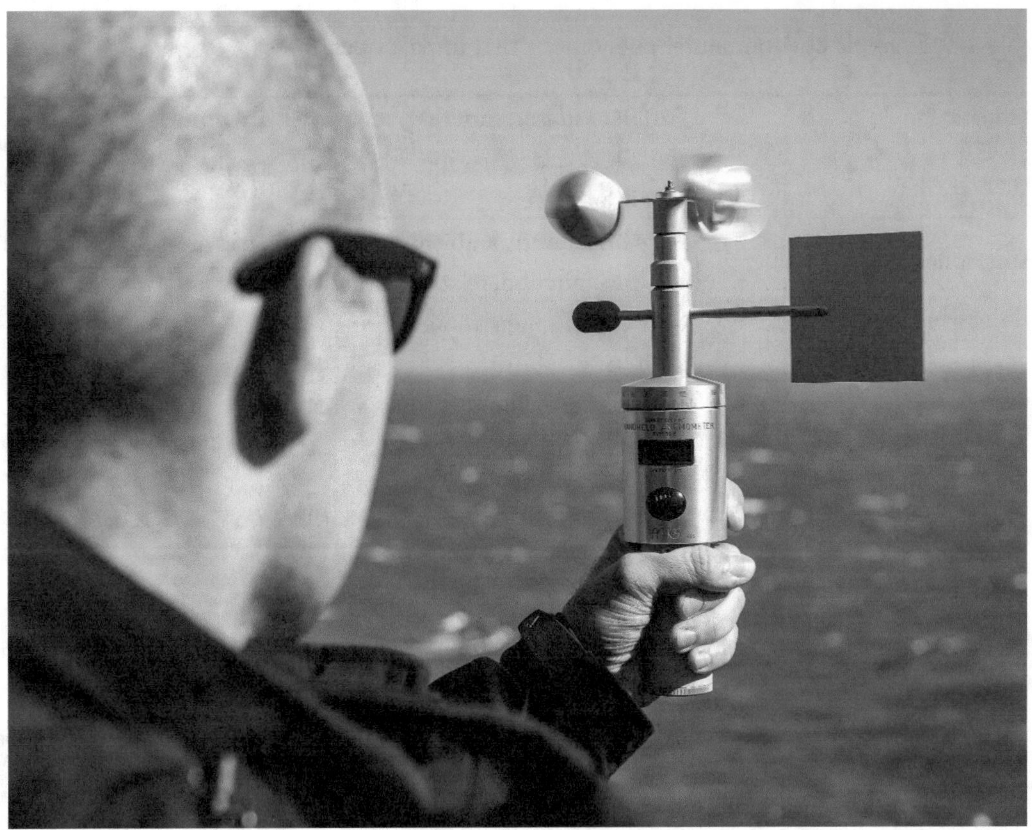

Figure 1.2: Measuring wind speed and direction with a handheld anemometer in support of maritime security operations in the Arabian Gulf (photo courtesy of the U.S. Navy).

ships (merchant, navy, research). ICOADS has been useful to naval architects in the testing of vessels, meteorologists, climatologists studying hurricanes, researchers assessing numerical model skill, scientists studying marine pollution, and the validation and calibration of satellite imagery [Berry and Kent, 2009, Wang et al., 2008, WMO, 2018].

Characteristics of important phenomena that impact marine structures and operations include wind, waves, currents, and water level fluctuations and these may be understood through information that is extracted from databases or discovered through online searches. Other variables for consideration include temperature, salinity, and biofouling. Biofouling is the undesired growth and accumulation of organisms on surfaces exposed to fluids and is a costly problem for many industries—from barnacles on weather buoy hulls to bacterial slime on transducers. Lobe et al. [2010] and Lobe [2015] have described methods to protect oceanographic instruments from the type of biofouling (stalked barnacles or gooseneck barnacles) that is shown in Fig. 1.3.

Table 1.1: Example environmental phenomena that impact operations

Acoustic	Ambient noise, scattering, sound velocity profile
Anthropogenic	Diffusers, jetties, marine debris, moorings, noise, outfalls, pollution, towers
Atmospheric	Air temperature, ambient light, barometric pressure, clouds, fog, precipitation, relative humidity, wind speed, wind velocity
Bathymetric and Topographic	Shoreline configuration, beach gradient, bottom type, reefs, sand waves, shifting channels
Biologic	Biofouling, bioluminescence, species abundance
Oceanographic	Acidity, depth, dissolved oxygen, heat flux, internal waves, salinity, sea state, surf, tides, turbidity, water temperature, wave direction, wave height

Table 1.2: Example meteorological and oceanographic sensor requirements for organizations such as the U.S. National Ocean Service and National Data Buoy Center

Parameter	Typical Reporting Range	Reporting Resolution	Sample Interval and Period	Accuracy
Air Temperature	-40°C to +45°C	0.1°C	10 min–avg, 10 min	± 0.1 K
Atmospheric Pressure	850 to 1,100 hPa	0.1 hPa	1 min–avg, 10 min	± 1 hPa
Dew point	-60°C to +45°C	N/A	10 min–avg, 10 min	± 0.1 K
Peak Wind Direction	0–359°	1°	3 sec–avg, 10 min	± 10°
Peak Wind Speed	0–95 m/s	0.1 m/s	3 sec–avg, 10 min	± 1 m/s
Relative Humidity	0–100%	0.1%	10 min–avg, 10 min	± 3%
Sea Surface Temperature	-4°C to +40°C	0.1°C	10 min–avg, 1 hr	± 0.1 K
Visibility	5–75,000 m	5% at 1,500 m	6 min	± 10%
Water Current	0–3 m/s	0.1 cm/s; 0.01° heading	6 min	± 0.25 cm/s
Water level	0–30 m	0.001 m	6 min	± 0.05 m
Wind Direction	0–359°	1°	10 min–avg, 10 min	± 10°
Wind Speed	0–95 m/s	0.1 m/s	10 min–avg, 10 min	± 1 m/s

Table 1.3: Dimensions and metric units for commonly used meteorological and oceanographic parameters

Quantity	Dimension	CGS Units	MKS Units
Acceleration	$\frac{L}{T^2}$	$\frac{\text{cm}}{\text{s}^2}$	$\frac{\text{m}}{\text{s}^2}$
Density	$\frac{L}{M^3}$	$\frac{\text{g}}{\text{cm}^3}$	$\frac{\text{kg}}{\text{m}^3}$
Force	$\frac{ML}{T^2}$	$\frac{\text{g cm}}{\text{s}^2}$	$\frac{\text{kg m}}{\text{s}^2}$
Length	L	cm	m
Mass	M	g	kg
Pressure	$\frac{M}{LT^2}$	$\frac{\text{g}}{\text{cm s}^2}$	$\frac{\text{kg}}{\text{ms}^2}$
Specific volume	$\frac{L^3}{M}$	$\frac{\text{cm}^3}{\text{g}}$	$\frac{\text{m}^3}{\text{kg}}$
Temperature	T	°C	°C
Time	T	s	s
Velocity	$\frac{L}{T}$	$\frac{\text{cm}}{\text{s}}$	$\frac{\text{m}}{\text{s}}$
Work	$\frac{ML^2}{T^2}$	$\frac{\text{g cm}^2}{\text{s}^2}$	$\frac{\text{kg m}^2}{\text{s}^2}$

Data centers ensure that information is discoverable, quality controlled, preserved, and presented in a high-quality way, and made available to the largest number of people. A national data center allows scientists and engineers to deposit data sets they have created in the hope that it can be repurposed. As described by the National Reseach Council [2003], government data centers have also been collecting data for decades and these data allow investigators to compare changing conditions through time and between locations. As a dedicated, central location for environmental datasets, the National Oceanic and Atmospheric Administration's (NOAA) National Center for Environmental Information (NCEI) in the United States ensures that disparate data collections are quality checked, archived daily and highly visible. However, comprehensive coverage for coastal and oceanic regions remains an aspiration rather than an achievement. Some researchers also fail to deposit their work into a data center, even when their research is funded through government grants. Data centers provide a starting point for any environmental characterization effort. Organizations such as NCEI bring consistency to the

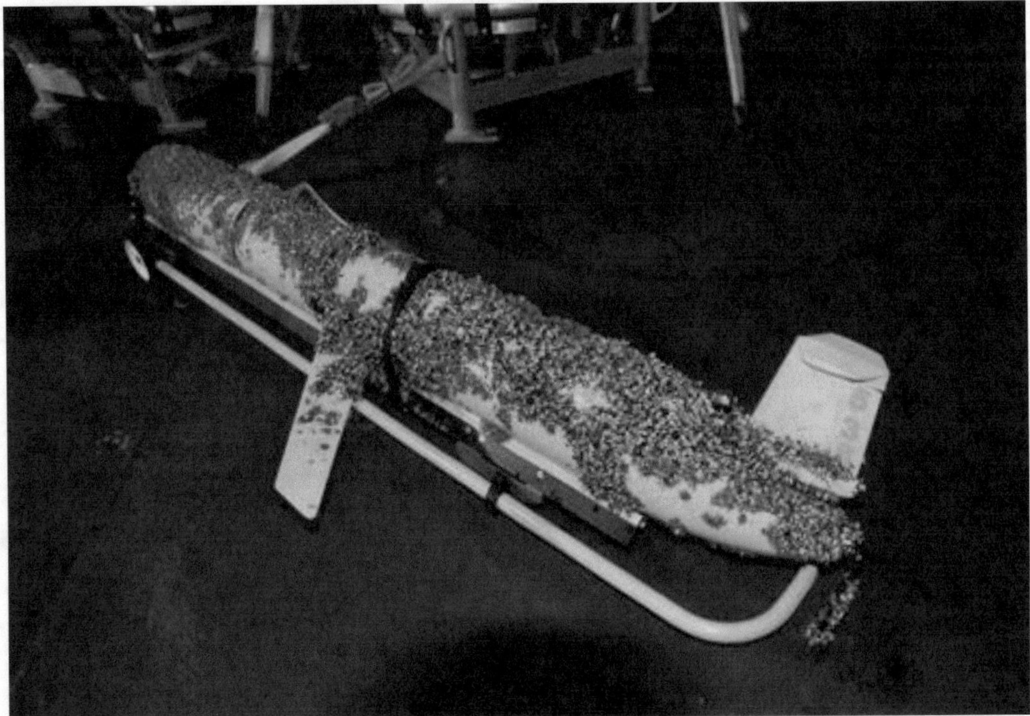

Figure 1.3: Biofouling on an Ocean Observatories Initiative (OOI) ocean glider recovered by the Research Vessel/Ice Breaker *Nathaniel B. Palmer* off the coast of Argentina in the South Atlantic on May 2, 2016 [Bigorre et al., 2017]. The Global Argentine Basin Array of the OOI combines surface moorings, subsurface moorings, and ocean gliders to provide air-sea fluxes of heat, moisture and momentum, and physical, biological and chemical properties throughout the water column. The accumulation of biological materials on meteorological and oceanographic instruments will affect different systems in different ways—a glider's response will be affected by the added mass of the biofouling.

data storage dilemma that faces many scientists and engineers. Metadata that document how observations were collected and the quality assurance/quality control protocols that were used are often provided. More recently, the NOAA sponsored U.S. Mesonet (Mesoscale Network) program supported the collection and archiving of non-federal asset data that is uploaded into the Mesonet archive every five minutes. Coastal atmospheric and oceanic Mesonet data collection sites are slowly becoming more prevalent and available, beginning the coverage for huge regional and national gaps. It is of note, that the U.S. National Weather Service (NWS) national network of Dual-Polarization (Dual-Pol) Radars collect 40,000 data points per second, 24 hours per day, 7 days per week, and 365 day per year, and have been in place for up to 7 years.

The raw radar data are available on-line from the NCEI. Additionally, Single Pol Radar data from the same NWS Radars are available back to 1988. This means that, collectively, there is a petabyte of meteorological data residing in the NCEI archives.

Remote sensing provides information on the extent, type, and dynamics of a study area. The retrieved information from remote sensing data, especially over large areas, generally requires the application of automatic classification methods. The resulting imagery provides the scientist or engineer with a spatially extensive view of their region of interest. Imagery and data are routinely collected from NOAA and National Aeronautics and Space Administration (NASA) satellites and aircraft and may require the use of in situ data for calibration and validation. Many nations are now sponsoring their own satellite systems with a vast array of advanced sensors. As an example, the Hyperspectral Imager for Coastal Oceans, which was tested aboard the International Space Station, and the Airborne Visible Infrared Imaging Spectrometer, which flies on NASA's high-altitude airborne science aircraft or ER-2, have provided valuable images that contributed to planning and executing humanitarian relief operations and locating oil spills [Zhao et al., 2018]. Data and image collecting drones are also becoming more commonplace, especially prior to, during, and following coastal storms. The U.S. National Science Foundation (NSF) has sponsored development of some of these types of modern technologies via its RAPID (Rapid Response Research) Program. Other research successes include implementation of a hyperspectral imaging capability and certification with the U.S. Federal Aviation Admnistration for the U.S. Navy's TigerShark Unmanned Aircraft System [Ball, 2019]. Figure 1.4 depicts a pair of Navy TigerShark aircraft, which are used for intelligence, surveillance, and reconnaissance missions. Georgia Tech Research Institute currently experiments with TigerShark by using payloads such as radar and other sensor modalities [Toon, 2016].

Data from a region of interest may be obtained by platforms such as ships, drifters, buoys, and moorings. World Meteorological Organization (WMO) Voluntary Observing Ship (VOS) programs provide carefully and rigorously collected meteorological and oceanographic data that are made available to users through resources such as ICOADS. The in situ data come in very different forms, from a single variable measured at a single point to multivariate four-dimensional collections of data that represent data volumes from a few bytes to gigabytes to petabytes. Data telemetry, which may rely on cellular or satellite communications, provides access to sensors such as floats that are deployed in the ocean. Example capabilities include Service Argos, which is being enhanced by the Kinéis nanosatellite constellation [Rogerson, 2019]. Sensors that comprise a NOAA Physical Oceanographic Real-Time System (PORTS ®) transmit data by radio to a data acquisition system that provides information products to a variety of users, including harbor pilots and tugboats. Nichols [1993] described the use of the Tampa Bay PORTS following Hurricane Andrew, which caused significant damage in South Florida during 1992.

Whenever possible, meteorological and oceanographic data are visualized over either topographic maps or bathymetric charts. Instruments including satellite tracked current drifters equipped with Global Positioning System (GPS) receivers and Acoustic Doppler Current Pro-

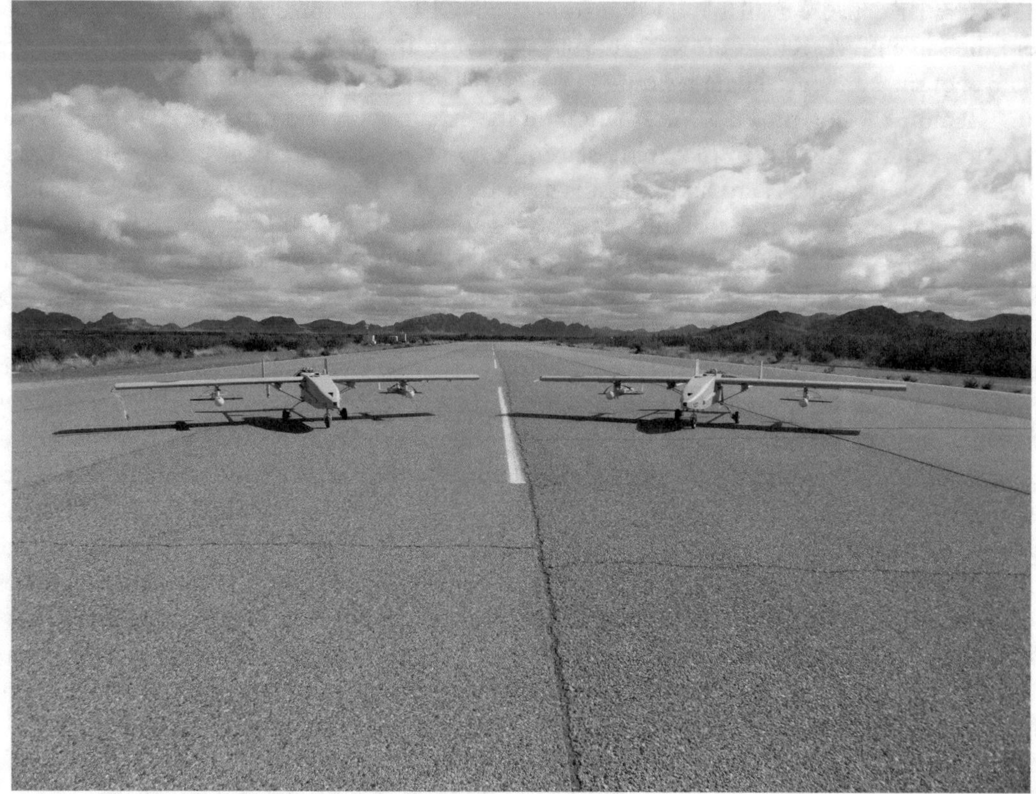

Figure 1.4: A pair of Navy TigerSharks on a runway during October 2018 carrying air-deployed Remora Unmanned Aerial Vehicles during Project Sea Krait. (Photo courtesy of the U.S. Navy.)

Figure 1.5: A window-shade drogue (left panel) and surface velocity program drifter (right panel) being deployed from naval vessels (photos courtesy of the U.S. Navy). Lumpkin et al. [2017] discussed unique oceanographic properties that can be observed with these types of drifters.

filers have been critical in supporting maritime projects in the Gulf of Mexico such as oil exploration and production. Figure 1.5 depicts several drifters that support maritime operations. Geospatial data collected from meteorological and oceanographic instruments are also important in the prediction of extreme conditions, especially for designers and operators.

Bathymetric surveys may be conducted to support operations such as charting, dredging, offshore drilling, and numerical modeling. Hydrographic surveyors and nautical cartographers tend to use multibeam sonar and a side scan sonar to understand seafloor characteristics while updating and maintaining nautical charts. Multibeam sonar measures the depth of the sea floor while the side scan identifies items on the seafloor. A single-beam depth finder might be used to determine depths directly under the vessel. Marine construction operations may include mandated bathymetric surveys to include underwater imagery. Requirements might be met with a side scan sonar and a Remotely Operated Vehicle (ROV). For example, the Bureau of Ocean Energy Management (BOEM) requests selected operators in the United States to perform visual habitat surveys before and after operations. Video that characterizes bottom conditions (e.g., color, texture, debris, and fish aggregations) must be delivered to BOEM along with side scan mosaics. Fox et al. [2001] provide procedures for the conduct of a nearshore rocky reef survey with an ROV. A tethered drone which has been designed for marine and aquatic environments is depicted in Fig. 1.6.

During the early 20th century, the International Hydrographic Organization (IHO) began the process of developing standards for the accuracy of echo sounding. Standards such as the S44 [IHO, 2008] support charting worldwide and have contributed to the development of international products such as the General Bathymetric Chart of the Oceans or GEBCO (see https://www.gebco.net/). Recent advances in surveying include the use of Unmanned Surface Vehicles (USV) such as University of New Hampshire's Bathymetric Explorer and Navigator or BEN, which is a diesel-powered *C-Worker 4* manufactured by L3 Harris ASV. Figure 1.7 depicts July 2017 BEN survey results that were used to populate gaps in Channel Island National Marine Sanctuary charts, characterize seafloor habitat, and help inform management decisions [Ballard et al., 2018, Raineault et al., 2018, Raineault and Ballard, 2018]. Since vessels may have difficulties collecting acoustic data in shallow waters, remote sensing techniques have been applied to retrieve water depth from overhead imagery [Bachmann et al., 2009a, 2010a, Jawak et al., 2015, Klemas, 2011, Smith and Sandwell, 2004]. Properly executed bathymetric surveys not only provide topography of the seafloor but also alert scientists and engineers on the orientation of seafloor cables, extent of bridge scour, the integrity of moorings, and the location of marine debris. Comparison of bathymetric data at the same location but from different time periods provides a method to assess rates of accretion or erosion. Circulation and wave modelers incorporate bathymetric data into their model by interpolating measured bathymetry to the model's grid.

Numerical models are an increasingly important means for studying the ocean and atmosphere in a controlled manner. As mentioned, ocean models require bathymetric datasets

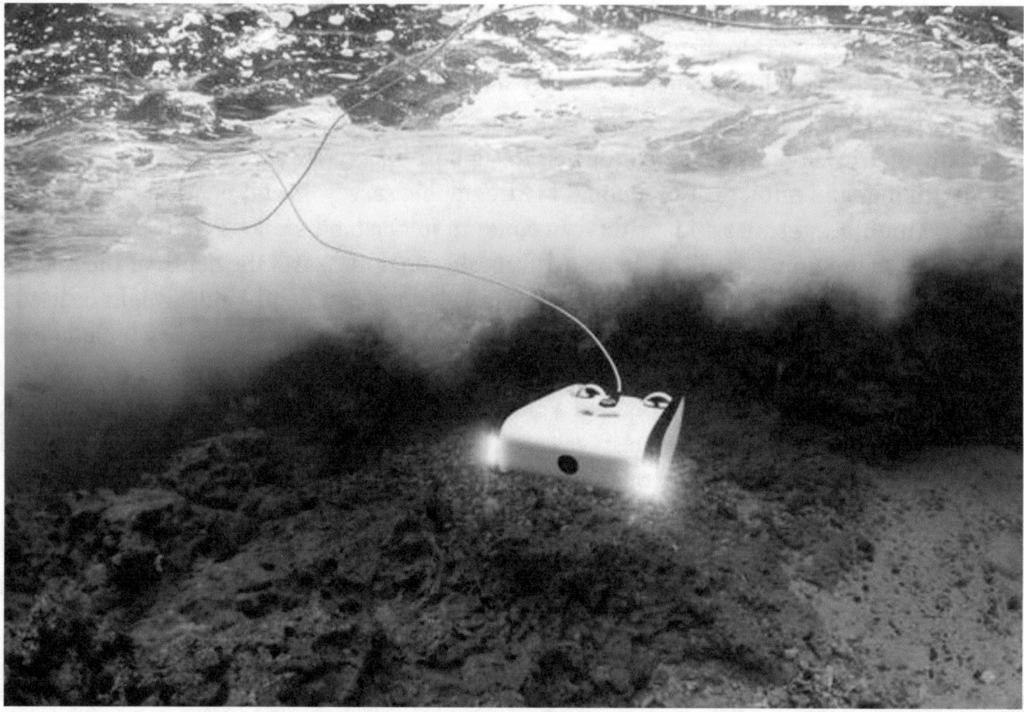

Figure 1.6: The Trident Underwater Drone in operation off Pupukea Beach, a lava-rock beach on the North Shore of Oahu in Hawaii. The Trident is ballasted with seawater, fitted with a 100 m tether and is designed to survey in coastal waters for three to four hours (photo courtesy of Domink Fretz and Sofar Ocean). Pupukea Beach (also known as Shark Cove) is a Marine Life Conservation District, an area designated by the Hawaii Department of Land and Natural Resources for the conservation and replenishment of marine resources.

to represent the seafloor. These bathymetric data may be derived from hydrographic surveys or satellite altimetry. Models incorporate field data to not only establish the model domain, but to also define boundary conditions such as waves, tides, water levels, and currents. Dynamical techniques in data assimilation, which include methods such as four-dimensional variational data assimilation, have involved the use of numerical models in conjunction with data made available from gliders, ocean observing systems, and satellite imagery [De Mey et al., 2009, Kurapov et al., 2011, Pasman et al., 2019, Song et al., 2016a,b]. Once validated, a numerical model can be used to simulate various design conditions to support operations. Model results that utilize statistical archived data are called hindcasts. Wave hindcasts that provide users, such as ocean engineers, with wave height, wave period, and wave direction are based on high-quality wind fields [Cox and Swail, 2001, Hanson et al., 2009]. Output that utilize real-time data to simu-

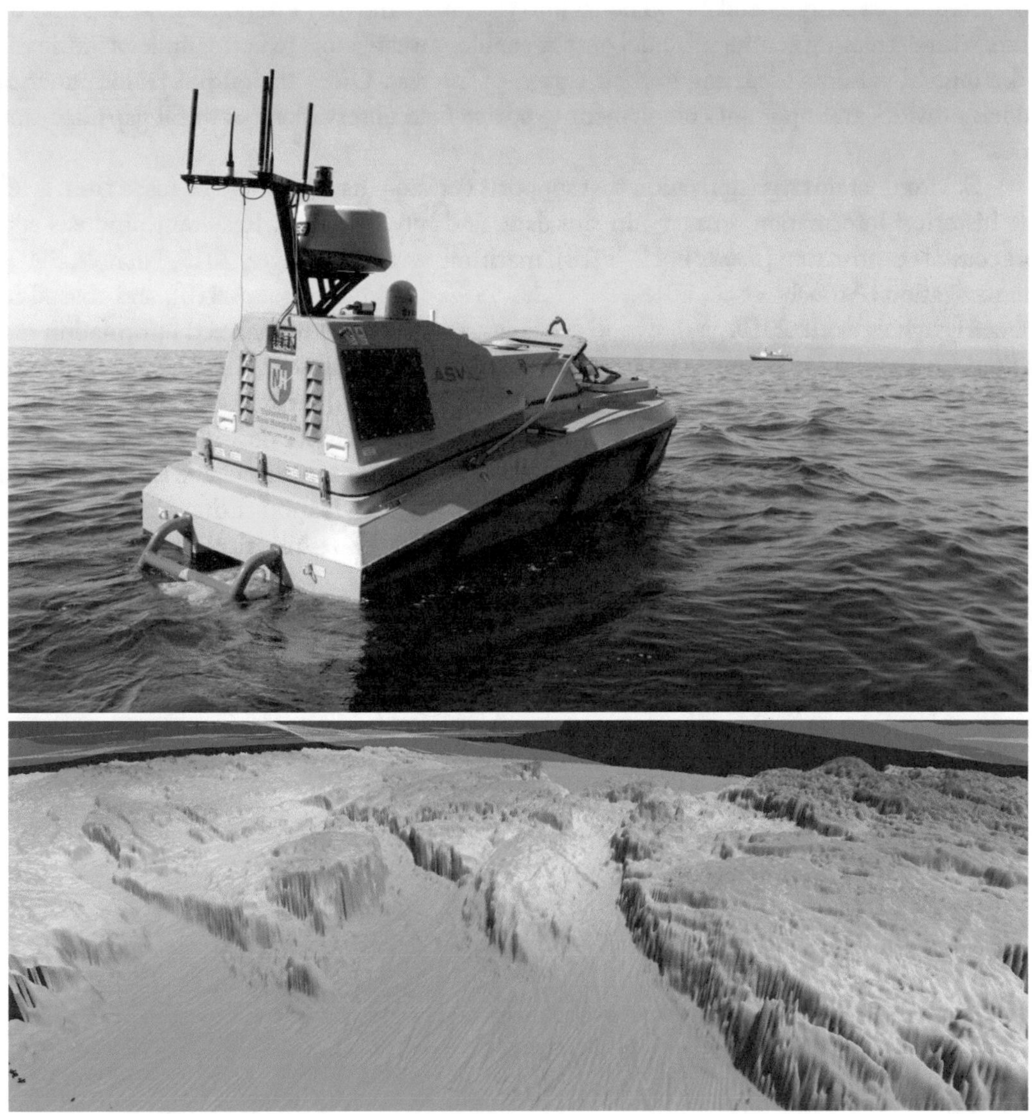

Figure 1.7: An instrumented USV with standard survey payloads was used to collect marine environmental data near Santa Barbara Island, CA during July 2017 (top panel, photo by Ed McNichol). Resultant multibeam bathymetry of the Channel Islands National Marine Sanctuary is presented in the bottom panel (images courtesy of the Ocean Exploration Trust and The Center for Coastal and Ocean Mapping at University of New Hampshire).

late present conditions are called nowcasts. Forecasts may be initialized from the most recent nowcasts and generally extend 24–48 hours into the future and may extend many weeks into the future with decreasing accuracy. Model output enables investigators to better understand spatial and temporal patterns occurring in their region of interest. Often the output from numerical models provides an important complement to sparse field observations or to fill gaps in a time series.

Environmental characterization that supports decision makers should be fused from available historical information, imagery, in situ data, and model output. Resultant products support coastal construction [Reeve et al., 2018], maritime security [Bueger, 2015, Nichols, 2003], safe navigation [Aroucha et al., 2018, Capelotti, 1996, Laughton et al., 2010], and coastal resilience [Nichols et al., 2019, Wright and Nichols, 2019]. Correctly analyzed information may be used for a variety of purposes such as risk analysis, the assessment of frequency responses, determining status and trends, computing environmental loads, verification and validation of forecast models, conducting trade studies, and making sustainability decisions. To support the consistency of information, this work cites several documents that are available from the Ocean Best Practices System repository, which is hosted by the United Nations Educational, Scientific and Cultural Organization (UNESCO; see https://www.oceanbestpractices.org/).

CHAPTER 2

Oceanographic Regions

The ocean is constantly changing as it absorbs heat from the sun and cools through sea to air evaporation and buoyancy transfers via heat fluxes, dissolves and releases carbon dioxide, generates waves in response to winds, and transports heat toward the poles from the tropics through Western Boundary Currents and Tropical and Extra-Tropical Cyclones. Such processes affect terrestrial environments, even those that are far from the coast. Scientists and engineers such as Cotton [1954], Fairbridge [2004], Finkl [2004], Inman and Nordstrom [1971], Shepard [1937], and Shepard [1973] have attempted to characterize regions based on physical and biological features and historic settings. In many cases, scientists and engineers may use Geographic Information Systems (GIS) to build category maps that allow users to visualize locations and the associated categories such as sea surface currents, heights, or temperatures.

In this text, we will focus on the coastal ocean, which is where ocean dynamics are modified by shallow water depth and the presence of land, thus affecting highly convective coastal frontal systems. In most situations, this means that the driving forces for water movement are the gravitational forces of the sun and moon upon the rotating earth and coastal zone wind stress. The wind stress changes with time and is the force per unit area that is affected by factors such as the wind speed and the shape of the wind waves. This region would include marginal seas and estuaries as well as nearshore and offshore regions. Of importance is the region between the nearshore and offshore, where factors that change rapidly, including light penetration, nutrient transport, and wave action have a strong influence on the ecology and impact important maritime operations. Data and information for these regions of interest are especially important for the determination of environmental loads.

2.1 OPEN OCEAN

The open ocean may be defined as the offshore extension of the continental margin, seaward of the continental shelves, which ends at the shelf-break zone, beyond which depths reach thousands of meters. It also includes shipping routes linking North America, Europe, and Pacific Asia through locations such as the Suez Canal, the Strait of Malacca, and the Panama Canal. Physical constraints on maritime operations such as shipping include winds, waves, currents, and sea ice. For example, energy transferred at the boundary between the atmosphere and the ocean by the wind and air pressure causes undulations of the sea surface, as indicated in Fig. 2.1. In the area of wave generation, the irregular waves that are growing are called "sea." The resulting surface waves propagating away from the area where they were generated are called "swell." Thus,

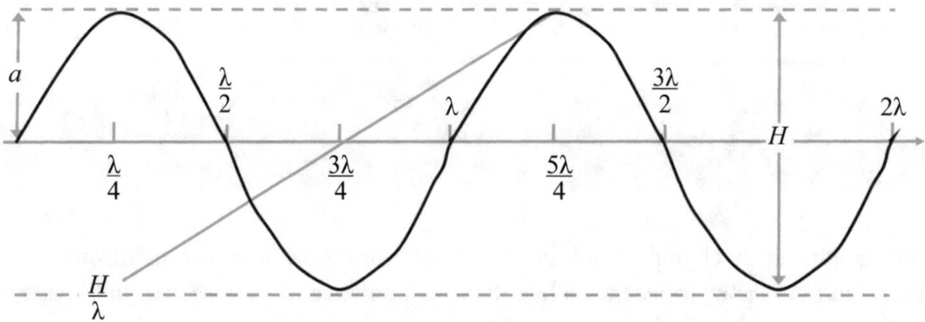

Figure 2.1: Amplitude, wavelength, height, period, and frequency can be used to study ocean waves. This graph represents an idealized ocean wave of sinusoidal shape. Information on waves are traditionally obtained from wave buoys or spectral wave models. Ocean gravity waves tend to have periods of 5–20 s, wavelengths of 100–200 m, and travel at speeds of 8–100 $\frac{\text{km}}{\text{hr}}$.

the energy delivered by the atmosphere to the ocean propagates toward the shore. Mariners are most concerned with wave heights and wave steepness, which impact ride quality and seakeeping. Wave height (H) is the distance from the trough to the peak in an idealized wave. The wave height is twice the amplitude (a). Wavelength (λ) is the distance between two successive points, such as crest to crest or trough to trough. Steepness, defined as the ratio of wave height divided by wavelength $(\frac{H}{\lambda})$, is the parameter that would most likely contribute to a maritime mishap such as capsizing. It generally decreases with increasing wavelength. The time interval between two successive points is the period (T) and is measured in seconds. The number of successive points passing a fixed point is the frequency (f) in Hertz.

The above parameters can be linked together to determine the celerity of propagation (wave velocity) as evidenced by the following general formula:

$$c = f\lambda, \tag{2.1}$$

where

$$\begin{aligned} c &= \text{ wave speed } \left(\tfrac{\text{m}}{\text{s}}\right), \\ f &= \text{ frequency } (\text{s}^{-1}), \text{ and} \\ \lambda &= \text{ wavelength (m)}. \end{aligned}$$

There are interrelationships among c, f, and λ. For example, wave period (T) is equal to (f^{-1}). Therefore, $c = \frac{\lambda}{T}$, i.e., the time taken for one wavelength to pass a fixed point. Further, wave number $(k) = \frac{2\pi}{\lambda}$ and the angular frequency $(\omega) = \frac{2\pi}{T}$. Wave number refers to the number of waves per meter while angular frequency refers to the number of waves (cycles) per second. This allows one to compute the celerity of propagation as angular frequency (radians$/s$) divided by wave number (radians$/m$) or $\frac{\omega}{k}(\frac{m}{s})$.

2.1.1 WAVES

Wave growth is known to be dependent on the speed of the wind, the duration of the wind, and the distance over which the wind blows or fetch. To first-order, wind-driven waves are associated with water particles that move in a circular orbit, so that there is practically no net movement of the water as the wave propagates out of the generating area. This circular motion diminishes exponentially with depth to the level of frictional influence. The motion of these waves may be characterized by a dispersion relationship which relates the group velocity to the wavenumber:

$$c^2 = \left(\frac{g}{k}\right) \tanh(kh), \tag{2.2}$$

where

$k = \frac{2\pi}{\lambda}$,
$h =$ water depth,
$g =$ acceleration due to gravity $\left(9.8\frac{\mathrm{m}}{\mathrm{s}^2}\right)$, and
$\lambda =$ wavelength.

This reduces to

$$c = \sqrt{\frac{g}{k}} \tag{2.3}$$

for deep water waves where the ratio of water depth to wavelength $\left(\frac{h}{\lambda}\right)$ is greater than 0.5. Thus, wave speed in deep water depends on wavelength and longer waves will travel faster than shorter waves. These waves show dispersion.

2.1.2 TIDES

In the ocean, astronomical and non-astronomical factors affect the height and timing of sea level. The periodic tides are caused by the gravitational forces between the Earth, Moon, and Sun. Since the moon is closer to the Earth than Sun, it exerts a stronger gravitational pull than does that of the sun. The Moon and Sun both create bulges of water on the side of the Earth that faces the Moon and separately the Sun, while the centrifugal force from the Earth's rotation causes bulges to form on the other side of the Earth. High tide occurs where these bulges occur. The sun's gravitational pull is 42% of that of the moon. Open-ocean tides and the ensuing currents are important in mixing deep-ocean water. Earth-orbiting satellites with radar altimeters are also able to measure tides over the deep-ocean. For example, U.S.-European satellite missions such as Jason-3, utilize a radar-altimeter to bounce microwave signals off the sea surface and precisely measures sea level [Leuliette and Nerem, 2016, Morrow and Le Traon, 2012]. Jason-3 is the result of a four-agency international partnership consisting of NOAA, NASA, the French Space Agency CNES (Center National d'Etudes Spatiales), and EUMETSAT (the European Organization for the Exploitation of Meteorological Satellites). With altimeter data, it is possible to determine what happens to tide waves and their energy as they travel across

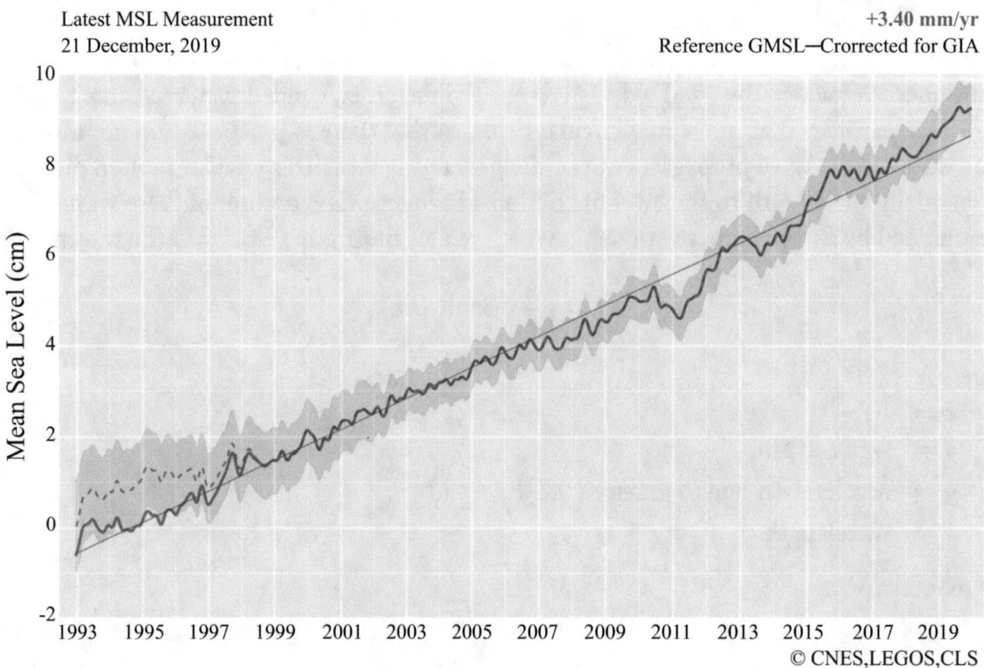

Figure 2.2: The global mean sea level rise (GMSL) rate from 1993–2019 was 3.40 mm/year (credits CNES, LEGOS, CLS). Glacial Isostatic Adjustment (GIA) accounts for the fact that the ocean basins are getting slightly larger since the end of the last glacial cycle.

deep ocean basins. Further, global sea level rise trends are calculated from satellite altimeters, as illustrated in Fig. 2.2.

Non-astronomical factors that affect height and timing of sea level include bathymetry, the speed of permanent currents, atmospheric pressure (the so-called "inverted barometer effect"), and sustained high winds from a constant direction. For example, during periods of high pressure, the water level tends to be lower than normal, and during periods of low pressure, the water level tends to be higher than normal. If there was a 1 mb change in atmospheric pressure, there would be an approximately 1 cm change in sea level. Near the coast, phenomena such as land and sea breezes develop owing to differential heating of air over land and sea during the spring and summer months. As illustrated in Fig. 2.3, sea breezes blow from the ocean inland toward land while the land breeze blows from the land to the ocean. Weather buoys and inland weather stations along locations such as the Florida Panhandle (also West Florida and Northwest Florida) provide data that are used to forecast afternoon thunderstorms caused by converging sea breezes.

Ocean currents result from forces such as the wind, the Coriolis effect, the Moon's orbit of the Earth and in turn, the Earth's orbit about the Sun, and temperature and salinity differences.

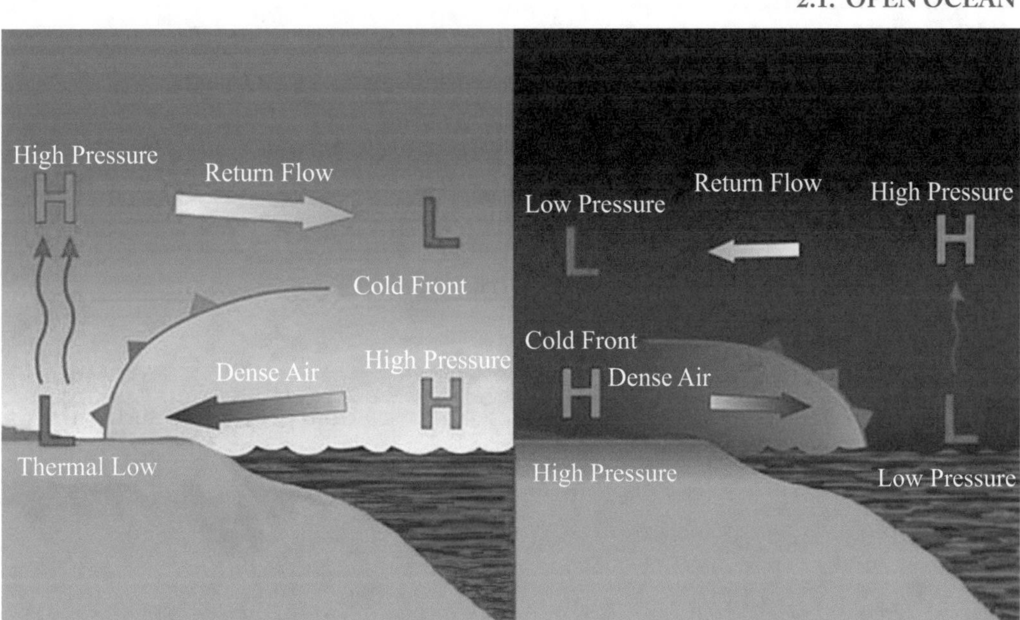

(a) Sea Breeze (b) Land Breeze

Figure 2.3: Sea- and land-breeze systems are set up by differential heating that is depicted by pressure gradients, cold front, and resulting (a) sea- and (b) land-breeze circulation (adapted from NOAA).

Once the wind sets surface waters in motion as a current, the Coriolis effect, Ekman transport, and the configuration of the ocean basin modulate the speed and direction of the current. Ocean Basin Western Boundary Currents, such as the Gulf Stream off the Atlantic Eastern Seaboard of the U.S., exist in all ocean basins, and are driven primarily by a balance between the horizontal pressure gradient and the Coriolis force. Tidal currents, which are caused by astronomical forces, change direction continuously in a rotary fashion in the open ocean. In a period of 12 hr and 25.2 min, the M_2 Principal Lunar tidal current will completely shift directions. There is no slack water as the current varies from hour to hour in an M_2 tidal day of 24 hr and 50 min. The Sun's Principal Tide (S_2), which has a 24-hr period, is experienced as a variation on the basic Lunar tide pattern, and not as a separate set of tides. Deep ocean currents are driven by density and temperature gradients. This thermohaline circulation is also known as the ocean's conveyor belt. The total current at a specific location is a combination of non-tidal and tidal currents and may be observed by current meters in the water column, drifters, and even radar. Current meters provide in situ measurement. Radar provides remotely sensed data for measurement of directional wave spectra and surface current. The current speed in a direction measured clockwise from geographic north is depicted in Fig. 2.4.

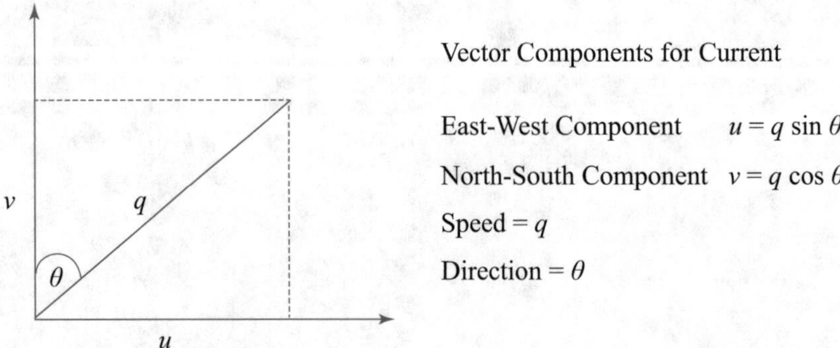

Figure 2.4: Current speed in a direction measured clockwise from geographic north. The total current is the sum of tidal and nontidal currents.

2.1.3 WATER QUALITY

There are many laws and policies that require the monitoring of environmental parameters for operations occurring on or near the ocean (e.g., U.S. Commision on Ocean Policy [2004]). Investigators may test water quality by sampling parameters such as salinity, temperature, dissolved oxygen, pH, and water clarity. Seawater is denser than both fresh water and pure water (density 1.0 kg/L at 4°C) because the dissolved salts increase the mass by a larger proportion than the volume. Much of the open ocean has a salinity ranging from 34–36 psu. Salinity is controlled by a balance between water removed by evaporation and freshwater added by rivers and rain. The Mediterranean Sea in Europe has very high salinity that is around 38 psu. It is almost closed off from the Atlantic Ocean and, owing to climate and geography, there is more evaporation than rain or extra freshwater added from rivers. In the deep ocean, water temperature is about 3°C and salinity ranges between 34–35 psu. Ocean temperatures are an especially important factor because in addition to providing information on heat content, temperature influences other parameters that may alter the physical and chemical properties of water. For example, if the water is too warm, it has a reduced ability to hold oxygen. Adequate dissolved oxygen is necessary for good water quality. Records of ocean temperature variability provide evidence for long-term processes such as decadal climate variability, El Niño and La Niña cycles, and global warming. Temperature and salinity also allow the calculation of water density, which can be used to understand dynamics such as stratification and flow directions. Figure 2.5 depicts the local collection of temperature and salinity profiles for a remote sensing study in Queensland, Australia, using a handheld castable instrument that provides instantaneous profiles of temperature, salinity, and sound speed [Bachmann et al., 2012a]. Since salinity and water temperature are closely related, one might characterize an area using a temperature salinity diagram. Salinity and temperature measurements may also be used to make sound velocity corrections to improve the accuracy of

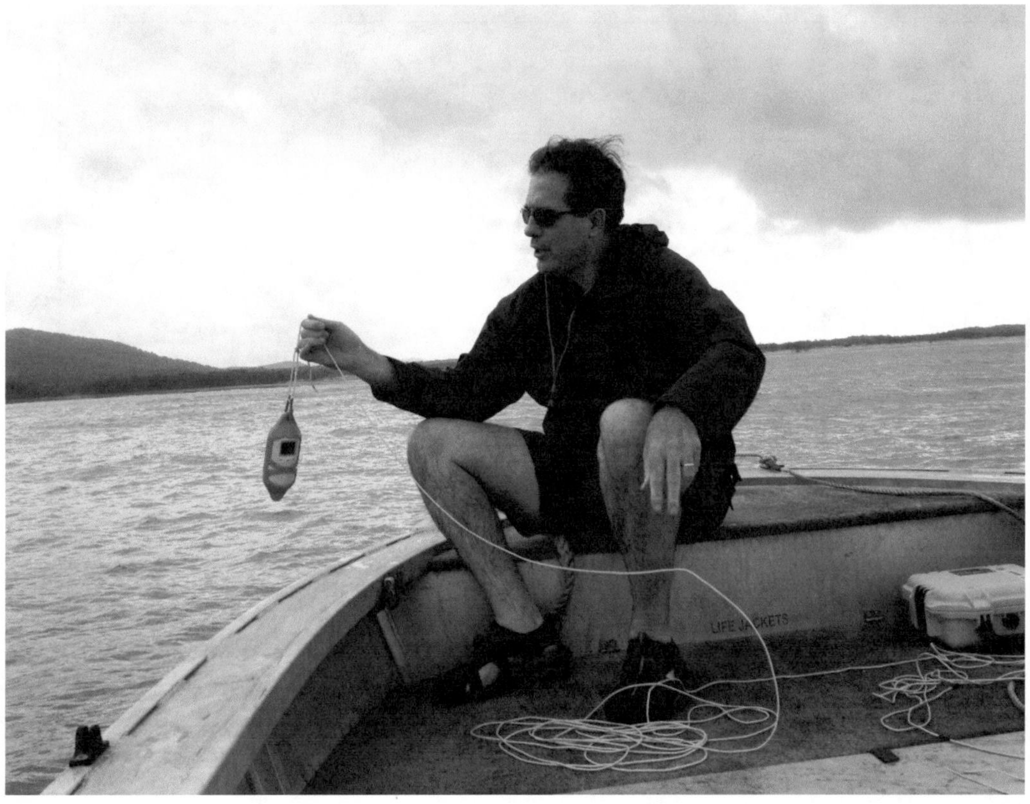

Figure 2.5: Collection of temperature and salinity profiles in coastal waters off of Queensland, Australia (photo by Deric Gray).

acoustically derived water level or current measurements. An illustration of global temperature salinity and density is provided in Fig. 2.6.

2.1.4 ANTHROPOGENIC IMPACTS

The majority of pollutants going into the ocean come from activities on land [Beiras, 2018]. Natural processes and human activities along the coastlines and far inland affect the health of our ocean. Nutrient pollution (nitrogen and phosphorous) contributes to harmful algal blooms (e.g., red tides, blue-green algae, and cyanobacteria) that can have severe impacts on human health, aquatic ecosystems, and the economy. Nonpoint source pollution occurs as a result of runoff and includes spillage from septic tanks, cars, trucks, and boats. Point source pollution comes from a single source such as during the Persian Gulf War Oil Spill, Deepwater Horizon Oil Spill, or discharge from faulty or damaged water treatment systems. Run-off containing nitrates and phosphates will lead to eutrophication and the ensuing reduced oxygen levels have been shown

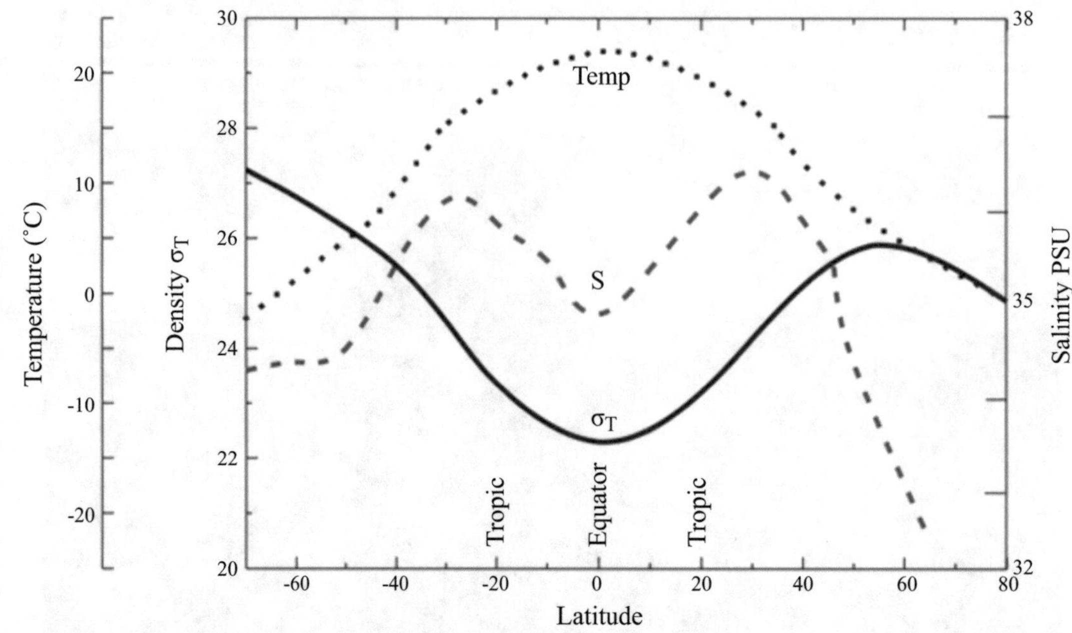

Figure 2.6: Global variation in surface temperature, salinity, and density [Pickard and Emery, 1982]. NASA provides online visualizations of sea surface temperature, salinity, and density at https://svs.gsfc.nasa.gov/3652.

to cause hypoxia or dead zones [Dybas, 2005, Joyce, 2000]. Examples of pollution from maritime activities includes ballast discharge from vessels and trash and debris from offshore oil and gas operations. Since ballast water typically contains a variety of biological materials, including plants, animals, viruses, and bacteria, it has contributed to the spread of invasive species. Hyun et al. [2017] demonstrated the impact of environmental factors such a nutrient concentrations and water temperatures on the survival of potentially invasive phytoplankton from ballast water.

Air pollution impacts water quality as it settles into waterways and oceans. Air pollution may be characterized in accordance to size, source, or to some sort of solid or liquid state. A simple classification of air particulates by size is depicted in Table 2.1. The U.S. Environmental Protection Agency has set National Ambient Air Quality Standards based on the mass concentration of "fine" particulate matter, which can be measured with instruments such as Aerodynamic Particle Sizers [Amaral et al., 2015]. Researchers Corbett and Fischbeck [1997, 2000], and Viana et al. [2014] have described how emissions from diesel powered vessels contribute air pollutants such as nitrogen oxides (NOx), sulphur oxide (SOx), fine particulate matter, hydrocarbons, carbon monoxide (CO), and carbon dioxide (CO_2). Vessel Automatic Identification System (AIS) datasets and environmental information should be integrated to more accurately estimate exhaust emissions from ships. AIS is a tracking system using transponders aboard ves-

Table 2.1: Particulates in air

Diameter less than .001 mm		Diameter greater than .001 mm
Aerosols	May be solid or liquid depending on their origin	Dusts (solid particles)
Smokes		Mists (liquid particles)
Fumes		

sels. These data may be exploited by vessel traffic and information systems. Integrated AIS and environmental information will help to produce more accurate information regarding ship exhaust emissions.

Water pollution is also caused by marine spills and the dumping of waste into the ocean by vessels. Ocean permanent currents that form gyres collect marine debris ranging in size from microplastics (< 5 mm in diameter) to large fishing nets and even shipping containers. Over the decades these collections of marine debris have formed extensive garbage patches [Debroas et al., 2017, Lebreton et al., 2018].

2.2 OFFSHORE

Physical constraints for operations such as shipping includes coastal configuration, winds, marine currents, depth, reefs, and sea ice. The offshore zone is not as well defined, but usually is described as the region beyond the breaker zone to the edge of the continental shelf. This region is important for shipping, commercial fishing, oil exploration, and production. Climate and the geographic location of marginal seas and passages are also critical for maritime operations. For example, narrow channels that connect two bodies of water along widely used sea routes are called "chokepoints." Hazards that potentially disrupt these chokepoints could have devastating effects on the transportation of supplies and put millions of people at risk. A list of important primary and secondary chokepoints is provided in Table 2.2.

The word "tides" is a generic term used to define the ocean's response principally to the gravitational attraction of the moon and the sun. Tides are the rise and fall of sea levels; this vertical motion of the water causes tidal currents, which are the corresponding horizontal motions of the water. Tides contribute greatly to variability in velocity, density, and pressure. *Tidalists* such as Doodson and Warburg [1941], Hicks [2006], Parker [2007], Shureman [1958], Talke and Jay [2013], Zervas [1999], and Zetler [1982] have described classical harmonic analysis techniques that can be used to harmonically predict tides and tidal currents. The primary international authority on the collection, quality control, and archival of tide gauge sea level observations is the Global Sea Level Observing System (GLOSS; http://www.gloss-sealevel.org), which is an international program established by the WMO and the UNESCO Intergovernmental Oceanographic Commission (IOC).

Table 2.2: Primary and secondary choke points supporting maritime commerce

Feature	Importance	Attribute
Dover Strait	Secondary	Connects the English Channel to the North Sea
Magellan Passage	Secondary	Passage between the Atlantic and Pacific Oceans
Panama Canal	Primary	Manmade waterway connecting the Atlantic and Pacific Oceans
Suez Canal	Primary	Manmade waterway connecting Mediterranean and Red Seas
Sunda Strait	Secondary	Connects the Java Sea to the Indian Ocean
Strait of Hormuz	Primary	Connects the Persian Gulf to the Gulf of Oman
Strait of Malacca	Primary	Connects the Andaman and South China Seas
Taiwan Strait	Secondary	Connects the East and South China Seas
These locations are key in the global trade of goods and commodities		

Tide and tidal current predictions are based on harmonic constituents which, in turn, are based on previous observations and do not take into account localized meteorological effects. Water level and current observations are usually collected and processed using software provided by the instrument manufacturer. Further, meteorological forces and variations in coastal configuration and bathymetry lead to tidal asymmetry, which is characterized by imbalanced falling and rising tidal periods and unequal peak flood and ebb currents. Real-time observation of water levels and currents may be more important to support operations owing to the differences in the effects that extreme weather can have on inshore and offshore water level fluctuations. While operations may not be conducted during severe weather, a storm surge can affect water levels for periods longer than the time of storm passage. Nichols [1993] described applications of the Tampa Bay PORTS during Hurricane Andrew which struck South Florida on August 24, 1992.

Defant [1961] described a classification scheme for tides based on the ratio formed by the amplitudes of two prominent diurnal constituents to two prominent semidiurnal constituents. This ratio is given as

$$\text{Form of tide} = \frac{K_1 + O_1}{M_2 + S_2}, \tag{2.4}$$

where

K_1 = Luni-solar declinational diurnal,
O_1 = Principal lunar declinational diurnal,
M_2 = Principal lunar semi-diurnal, and
S_2 = Principal solar semi-diurnal.

These ratios differ along large stretches of the coast and can be used to describe areas where the tidal characteristics are diurnal, semidiurnal, and mixed.

In the offshore, tidal currents may be characterized by an ellipse where the current vectors typically trace out a rotary motion over the course of a tidal cycle, as shown in Fig. 2.7. The size of the current ellipse is proportional to the current magnitude. Foreman [1978] described how tidal current ellipses may be written in the complex form as

$$u + iv = A^{+}e^{i\omega t} + A^{-}e^{-i\omega t}, \tag{2.5}$$

where

u = the east-west velocity,
v = the north-south velocity, and
ω = the angular frequency,

and complex vectors rotating in opposite directions that are given by

$$A^{+} = \text{counterclockwise vector, and}$$
$$A^{-} = \text{clockwise vector.}$$

Observed and modeled tidal ellipses may also be used to assess the skills of simulated tidal currents [Cummins and Pramod, 2018].

In offshore regions, ocean waves may transition from deep to shallow water. When the depth of the water, h, becomes less than one half of the wavelength, the progressive waves will no longer show dispersion. The shallow water wave is independent of wavelength. Wave speed depends on the depth of the water as indicated by

$$c = \sqrt{gh}, \tag{2.6}$$

where

g = acceleration due to gravity (9.8 $\frac{\text{m}}{\text{s}^2}$) and
h = water depth.

Waves in the ocean may also be formed by Earth process impacts such as undersea earthquakes and continental margin landslides, which impart energy to the ocean that is dynamically repackaged as very long, low amplitude propagating waves called tsunamis. These are shallow-water waves, since their wavelength may be longer than twice the ocean's depth. Tsunami may also be described by their period, amplitude, and speed. The amplitude of normal ocean waves and tsunami are similar in deep ocean waters, but as the tsunami propagate toward the shore, they slow down and their back ends catch up to their fronts and they grow in amplitude, with heights of 10 m or more. Tsunami tend to have much longer periods of 10 min to 2 hr, wavelengths of 100–500 km, and travel at speeds of 800–1000 $\frac{\text{km}}{\text{hr}}$, so can be very dangerous to coastal structures and populations as they converge rapidly onto and into a coast.

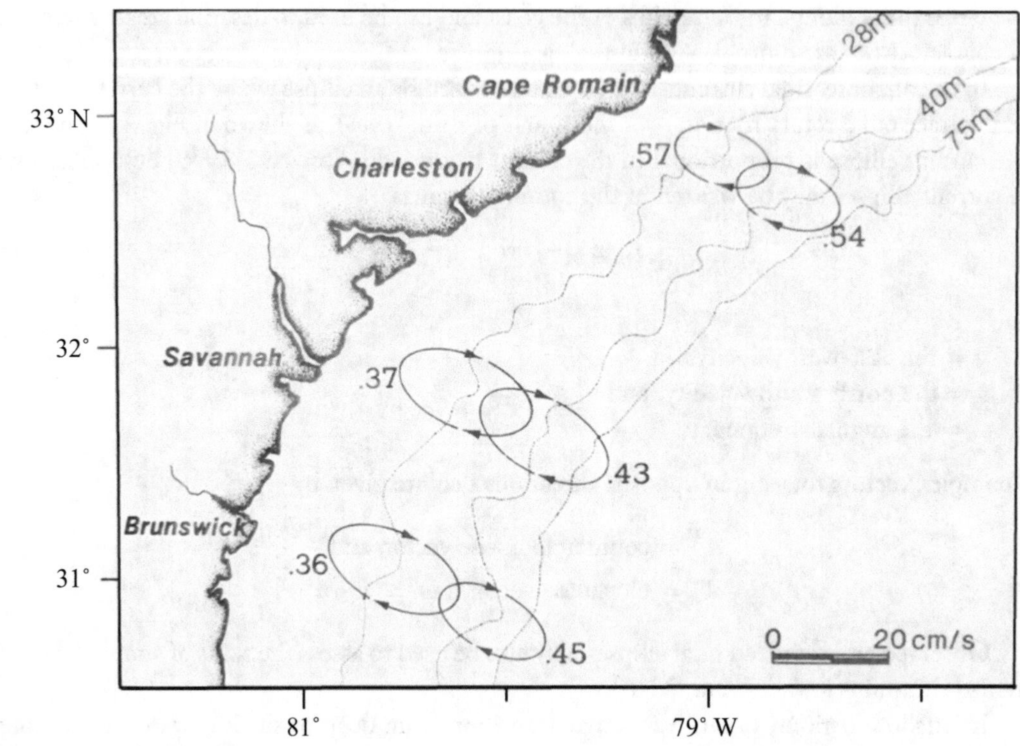

Figure 2.7: Surface tidal current ellipses for the principal lunar semidiurnal (M_2) component along the South Atlantic Bight [Pietrafesa et al., 1985]. Ellipticity is given as the ratio of v to u components.

2.3 NEARSHORE

Nearshore marine regions are physically contiguous to the open ocean. Along the coastal ocean, the nearshore generally extends from the low water line well beyond the surf zone where the area is influenced by longshore and rip currents. This region extends from the upper limit of a beach to the offshore. In terms of the beach profile, it consists of (progressing seawards) the backshore, foreshore and nearshore. In terms of current and wave regimes, the nearshore consists of the breaker zone, surf zone, and swash zone. Longshore and rip currents are common at any beach that is exposed to surf.

2.3.1 WAVES

In the breaker zone, waves become unstable and break as collapsing, spilling, plunging, or surging surf when maximum wave height is approximately .78 times the shallow water depth. The particular form of breaking waves is dependent on factors such as wind and beach gradients.

Research related to nearshore processes is conducted at locations such as the U.S. Army Corps of Engineers' Field Research Facility (FRF) in Duck, North Carolina. The FRF is recognized as one of the few nearshore observatories that provide access to historic and real-time observations and model output through a data portal (see https://frfdataportal.erdc.dren.mil/). At the FRF, the skill of wave models such as Simulating Waves Nearshore (SWAN), XBeach, and Delft3D, which use local bathymetry to transform the offshore wave conditions into shallow water, have been tested. These models incorporate wind data to simulate wave growth or decay and include the effects of currents and tides [Caldwell and Aucan, 2007, Hsu et al., 2006, 2010]. Team science experiments such as DELILAH, DUCK '94, and SandyDuck have been conducted at the FRF to better understand waves, tides, and shallow water processes. Xia et al. [2020] recently described how observed wave and current data collected from the DUCK '94 Experiment were used to validate a wave-current modeling system. The Finite-Volume Community Ocean Model (FVCOM) and Surface Wave Model (SWAVE) were coupled to better understand wave-induced circulation. Models such as SWAN are being applied by some NWS offices to improve safety at selected beaches in the United States through surf zone forecasts (see https://www.weather.gov/safety/ripcurrent-forecasts).

2.3.2 TIDES

When a tidal current moves toward the land and away from the open ocean, it "floods." When it moves toward the open ocean or away from the land, it "ebbs." These tidal currents that ebb and flood in opposite directions are called "rectilinear" or "reversing" currents. Rectilinear tidal currents, which typically are found in coastal rivers and estuaries, experience a "slack water" period of no velocity as they move from the ebbing to flooding stage and vice versa. After a brief slack period, which can range from seconds to several minutes and generally coincides with high or low tide, the current switches direction and increases in velocity. If a moored current meter is deployed, the rectilinear shape of the flow can be depicted by making a scatter plot of u and v velocity components.

Non-tidal forces may also affect height and timing of sea level. These include the impact of bathymetry, the land-sea breeze cycle, and sustained high winds from a constant direction which sets up the storm surge. For example, differences in water levels inside and outside of a coastal inlet set up a pressure head that drives hydraulic currents. The pressure head is observed by bottom-mounted pressure gages and the hydraulic currents are observed by current meters as exaggerated ebb and flood currents.

River mouths and coastal inlets also connect the coastal ocean to inland waterways. Important marine operations such as dredging are used to maintain navigation through these passages and requires an understanding of the various hydrodynamic and geomorphic processes that impact nearshore regions. These regions may be the location of critical habitat, e.g., some nearshore waters of the mid-Atlantic are important to endangered North Atlantic right whales *Eubalaena glacialis* [Whitt et al., 2013].

2.3.3 WATER QUALITY

The proximity to land and freshwater sources affects estuarine areas such as coastal bays, river mouths, and tidal marshes, as illustrated by salinity observations. Freshwater from land drainage and seawater combine to form brackish, or slightly salty water. Freshwater river plumes often extend offshore into nearshore coastal zones and, in the absence of wind forcing or an ambient long-shelf current, tend to turn right as they enter the coastal ocean in the northern hemisphere. Horner-Devine et al. [2015] reviewed plume dynamics and include research results that describe flows of the Chesapeake Bay, Columbia River, and Mississippi River plumes into the Atlantic Ocean, Pacific Ocean, and Gulf of Mexico, respectively.

As evidenced by damages to coastal structures such as piers during tsunami or tropical cyclones and the capsizing of boats by steep waves, shallow water waves have a significant impact on marine structures and safe navigation. Waves propagating on top of a storm surge compound the effects of inundation, which include property loss, dangers such as drownings, and the contamination of drinking water with microorganisms (bacteria, parasites, viruses) and toxic chemicals (cleaners, pesticides, petroleum). In order to ensure safe engineering designs and marine operations, important wave processes to consider are refraction, diffraction, reflection, wave breaking, wave-current interaction, friction, wave growth due to the wind, and wave shoaling. Such factors can be used to assess fatigue, compute loads, or to provide mariners with warnings that facilitate safe navigation.

Wave and current processes contribute to sea surface roughness and the transport of sediment laden water. Swell and sea may increase sun glint, which can degrade different types of imagery that are used to obtain spatially extensive information. Sun glint occurs when sunlight reflects off the sea surface at the same angle that a remote sensor is viewing the surface. Waves and currents influence the amount of suspended materials that impacts light attenuation and the health of marine organisms. As deep-water waves approach shore and feel the bottom, the wave height increases, wavelengths decrease, and the frequency remains unchanged. These shoaling effects are related to group velocity changes in shallow water. The group velocity of the waves is given by

$$v = \frac{c}{2}\left[1 - 2kh \operatorname{cosech}(kh)\right], \tag{2.7}$$

where

v = group velocity,
c = wave speed,
k = wave number, and
h = depth.

In deep water, the group velocity is equal to half of the celerity. As the waves shoal or move into shallow water they will eventually break as surf once wave heights are approximately .78 times the depth. Owing to factors such as wind and depth profiles, breakers will take the form of plunging, spilling, surging, or collapsing surf. Nearshore currents, shoaling waves, and surf contribute to increased turbidity owing to the resuspension of bottom materials such as silts, clays, and sand. Since particles in the water absorb and reflect light, spectral remote sensing is a very useful tool to estimate water quality and map plumes of increased turbidity caused by runoff, nearshore currents, or wave action.

2.3.4 ANTHROPOGENIC IMPACTS

Ocean monitoring is imperative to ensure the protection of human health and ecosystems from the harmful effects of bacterial contamination and pollution. Rainfall, tides, and storm surge have the potential to carry disease-causing bacteria to the beach and into the ocean. Many municipalities collect water samples near public beaches and post advisories when high numbers of bacteria are found. For example, *Enterococcus* strains may be tested for in the water samples. While *Enterococci* are typically not considered harmful to humans, their presence is associated with pathogens such as norovirus, *Shigella* and *Escherichia coli*. Bacteria concentrations may vary owing to factors such as tidal currents, sunlight (cloud cover), and El Niño events which cause increased rainfall.

Environmental characterization is especially important to quantify the increased intensity and frequency of storms and hurricanes, coupled with sea level rise, and changes in the land and seascape. Growing coastal populations and the corresponding increases in infrastructure places greater numbers of people and structures at risk for damage from coastal hazards (e.g., recurring floods, hurricane force winds, and storm surge). Wright and Nichols [2019] highlighted the impact of storms and floods on coastal communities, worldwide. To plan for these threats, the U.S. Army Corps of Engineers provides a web-based Sea-Level Change Curve Calculator that projects rates of sea level change for any location along the United States' coast based on data and information available through NOAA. The tool can be used to assess an area's vulnerability to future sea level change. The tool is available at http://corpsmapu.usace.army.mil/rccinfo/slc/slcc_calc.html. The U.S. Army Corps of Engineers Sea Level Change Curve Calculator uses the methodology described in Engineer Regulation No. 1100-2-8162 entitled, *Incorporating Sea Level Changes in Civil Works Programs* [U.S. Army Corps of Engineers, 2013]. NOAA has a similar Sea Level Rise Map Viewer that can be accessed at https://www.climate.gov/maps-data/dataset/sea-level-rise-map-viewer. An element of the United States' Strategic Environmental Development Program (SERDP) supports the application of science and technology advances to enhance the management of military installations through improved coastal resilience. Hall et al. [2016] discuss future sea level rise and local extreme water level events for military sites located around the globe. Kaus Raghukumar discussed risk analysis to improve resilience at the 2019 SERDP and ESTCP Symposium[1] [Jones et al., 2019]. Effective coastal resilience will require the integration of engineering, environmental science, and community-based approaches. Wright et al. [2016] identified the need

[1]2019 SERDP and ESTCP Symposium, https://serdp-estcp.org/News-and-Events/Conferences-Workshops/2019-Symposium/2019-Symposium-Archive.

for interdisciplinary scientists and a supporting cyberinfrastructure with emphasis on understanding and predicting the future behavior of coastal systems based on disparate types of data.

CHAPTER 3

Example Critical Phenomena

Environmental characteristics are usually described by physical variables in a statistical nature. The statistical description should reveal the extreme conditions as well as the long- and short-term variations. If a reliable simultaneous database exists, the environmental phenomena can be described by joint probabilities. For example, the joint distribution of wave height and direction might be used to characterize the wave field, where wave height is a scalar value and wave direction is an angular value. Such approaches may also be used to calculate the likelihood of waves and water levels occurring together and at the same point in time [Hawkes et al., 2002]. This would be especially important to evaluate the impact of waves on navigation and to assess flood risk along macrotidal coasts that are subject to surges.

Wave parameters are often used to classify the irregular sea. A common method involves fitting a cumulative probability distribution for significant wave height (H_S) to long-term field measurements, and the short-term distribution of sea state is assumed to follow a Rayleigh distribution, as indicated in Fig. 3.1. The average height (H_o) generally occurs less frequently than the height of the most probable wave. H_S is the arithmetic average of the highest one third of the waves in a wave record. H_S is associated with significant wave period (T_S). In some cases, H_S and T_S might be used to classify the irregular sea, but a more thorough classification would include parameters such as average wave period and maximum wave height ($H_{\max}$). $H_{\max}$, which occurs infrequently, is often estimated as Li et al. [2016]:

$$H_{\max} = 1.77 H_S. \tag{3.1}$$

Since extreme waves will cause the highest loads and potential for wave overtopping on structures, other methods may be used to ensure that there is no under-prediction of the design maximum wave height which could lead to unsafe designs or an over-prediction which could contribute to over-engineered and expensive designs. Rogue waves are defined as waves that have a height that is more than twice the significant wave height. On January 1, 1995, a destructive 25.6 m maximum wave height was recorded at the Draupner platform in the North Sea and this wave has recently been modeled by McAllister et al. [2019]. Statistical wave height distributions such as those described above may transform the significant wave height to lower exceedance wave heights. Significant wave parameters may also be obtained from buoys or numerical wave propagation models where design values can be based on extreme value analysis.

Scientists and engineers typically determine the potential for extreme conditions using historical data from nearby sites [Caires, 2011]. Observed wave climate is displayed as a joint frequency or scatter table by combining both wave height (e.g., significant wave height) and wave

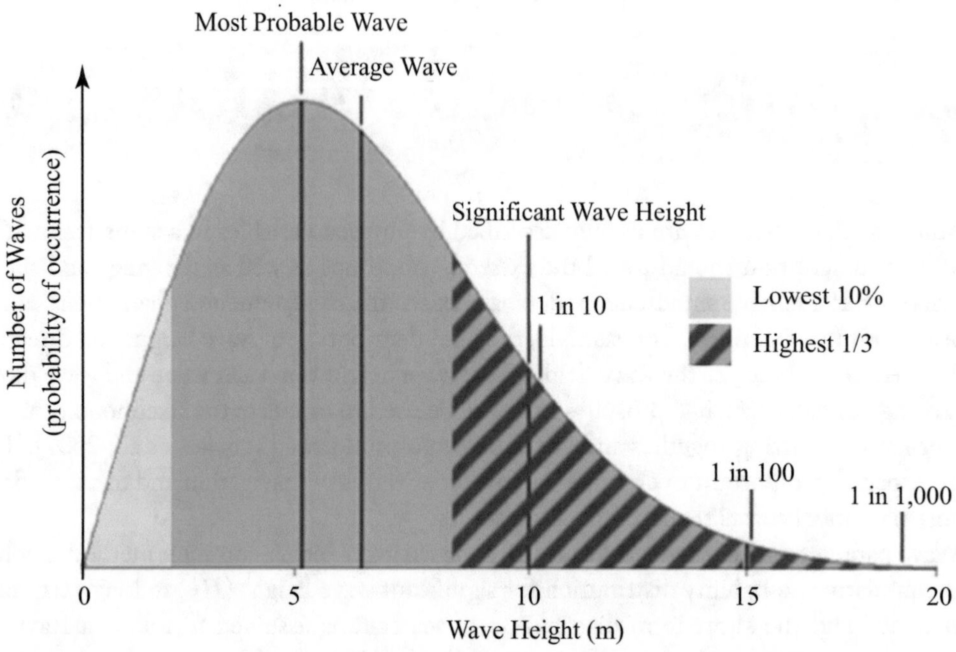

Figure 3.1: The Rayleigh distribution of wave heights. Longuet-Higgins [1952] showed that random wave heights, H, followed a Rayleigh Probability Distribution.

period (e.g., peak wave period) data in a selected region over a long time period. Wave climate studies usually use data from a combination of sources (e.g., ship observations, altimeters, buoys, and model output) to specify offshore conditions [Shope et al., 2016]. Long-term probability distributions that are used to calculate the 100-year wave height are based on the probability distribution of individual wave heights at a study site. One might also consider the long-term probability distribution of the extreme wave heights during a fixed period of time.

Empirical, statistical data used as a basis for evaluation of operation and design must cover a sufficiently long time period. For operations of a limited duration, seasonal variations must be taken into account. For meteorological and oceanographic data, several decades of observations should be available. If the data record is shorter, the climatic uncertainty should be included in the analysis. Williams and Esteves [2017] describe approaches to validate numerical wave models that can then be used to forecast average wave characteristics years into the future.

CHAPTER 4

Systems and Sensors

Ocean and coastal engineering projects that involve construction, repairs, and maintenance usually include inspections, site surveys, and design. Example projects are listed in Table 4.1. Site surveys will include the collection of data to characterize the environment. For example, understanding the trajectory and force of the Loop Current is crucial to oil production in the Gulf of Mexico. Various sensors will be required to collect data and the associated platforms that allow the sensors to be maintained, and communication technologies that transmit the data to a user or data collection center, often with satellite or cell phone telemetry. Platforms such as Autonomous Underwater Vehicles (AUVs) are frequently used to collect oceanographic data to support engineering and maritime surveillance projects (see Fig. 4.1). Table 4.2 lists some of the many types of measurements that can be collected to characterize the environment. These data are generally collected from platforms which accommodate instruments and in some cases scientists and engineers. The National Data Buoy Center (NDBC) of NOAA operates 45 Coastal-Marine Automatic Network (C-MAN) stations in harsh and often remote marine environments around the United States. A picture of a C-MAN station is provided in Fig. 4.2. Other example platforms are listed in Table 4.3.

Computer models that produce forecasts of ocean conditions are also especially important since it is not practical to have sensors everywhere, and the model output can fill in the voids. Model simulations may be forced by harmonically predicted tides, climatological data, or information (e.g., winds, water levels) from larger-scale models. Significant wave height output from the Wave Watch III model for the Pacific Ocean on March 31, 2020 are provided in Fig. 4.3. Wind input is from the operational Global Forecast System that is run by the NWS. Instrumented buoys also provide essential data for the development and testing of numerical models [Bidlot and Holt, 2006]. While the use of models to support safe and efficient marine navigation, emergency response, and ecological applications is valuable, skill assessments to ensure reliability requires models to meet or exceed target benchmarks before forecasts are approved for release to the public [Zhang et al., 2010]. The skill assessment should focus on variables, statistics, and targets that are important to the users.

Instruments and measurement techniques extend the ability of scientists and engineers to observe and characterize the environment. In marine science much is still to be learned through observation and the limit to observation is often instrumentation. Skilled marine scientists and technologists will deploy oceanographic instrumentation at optimal locations to generate greater and more accurate understanding of oceanographic systems. One advantage from improved un-

Table 4.1: Selected ocean and coastal engineering projects

Boat ramp construction
Bridge and dam inspection/repair
Concrete work (pier/piling repairs)
Cutting and welding
Demolition and obstacle removal
Dock, pier, and bulkhead construction
Dredging
Hydrographic survey and charting
Pipeline installation/repairs
Salvage
Seafloor cable installation/repairs
Shoreline protection
Unmanned underwater vehicle operations
Video/still photography

Table 4.2: Traditional measurements used to characterize the environment

Air and water temperature
Atmospheric pressure
Current speed and direction
Precipitation
Relative humidity
Salinity
Solar radiation
Underwater sound
Visibility
Water-column height
Water level and water quality
Wave energy spectra (non-directional and directional)
Wind direction, speed, and gust

Figure 4.1: A Dutch Navy team recovers an AUV while performing a bottom of the sea reconnaissance prior to amphibious maneuvers during NATO Exercise Trident Juncture 18 (NATO photo by WO FRAN C. Valverde).

derstanding of phenomena is the ability to predict ocean responses. Information about the various sensors and samplers oceanographers use to study the ocean are provided in Table 4.4.

Oceanographic platforms are installed with a variety of sensors. Traditional platforms to sample the ocean include tide houses, C-MAN stations, research vessels, moorings, autonomous underwater vehicles, unmanned surface vehicles, near-surface drifters, floats, aircraft, and satellites. Tide houses and C-MAN stations are structures that protect sensitive electronics, transmitting equipment, and backup power and data storage devices. Example sensors that are fixed to these structures include anemometers, pressure gages, and acoustic sounding tubes for water levels and possibly a backup pressure sensor. Data are transmitted back to an operational organization such as NOAA for analysis and distribution. In the United States more than 200 water level recorders comprise the National Water Level Observation Network (NWLON), which are safeguarded in tide houses. Some of these water level observations date back to the 1850s. High technology research vessels such as the Norwegian ship *GO Sars* have been designed to support research on the physical, chemical, and biological char-

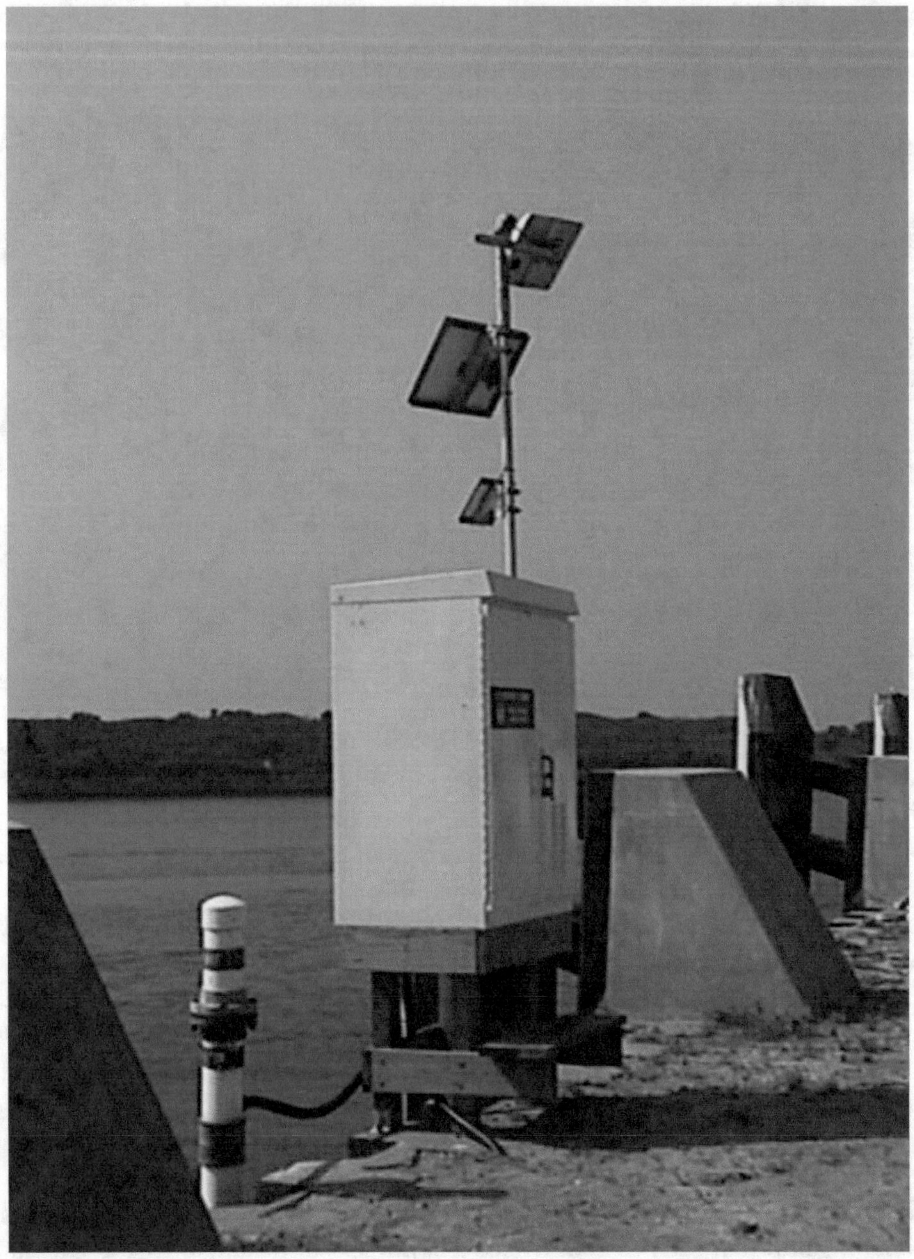

Figure 4.2: NDBC C-MAN Station 8536110 located near the Cape May Lewes Ferry Terminal on the Cape May Canal, which was built by the U.S. Army Corps of Engineers and first opened during 1942 (Dorwart [1992]; photo obtained from NOAA).

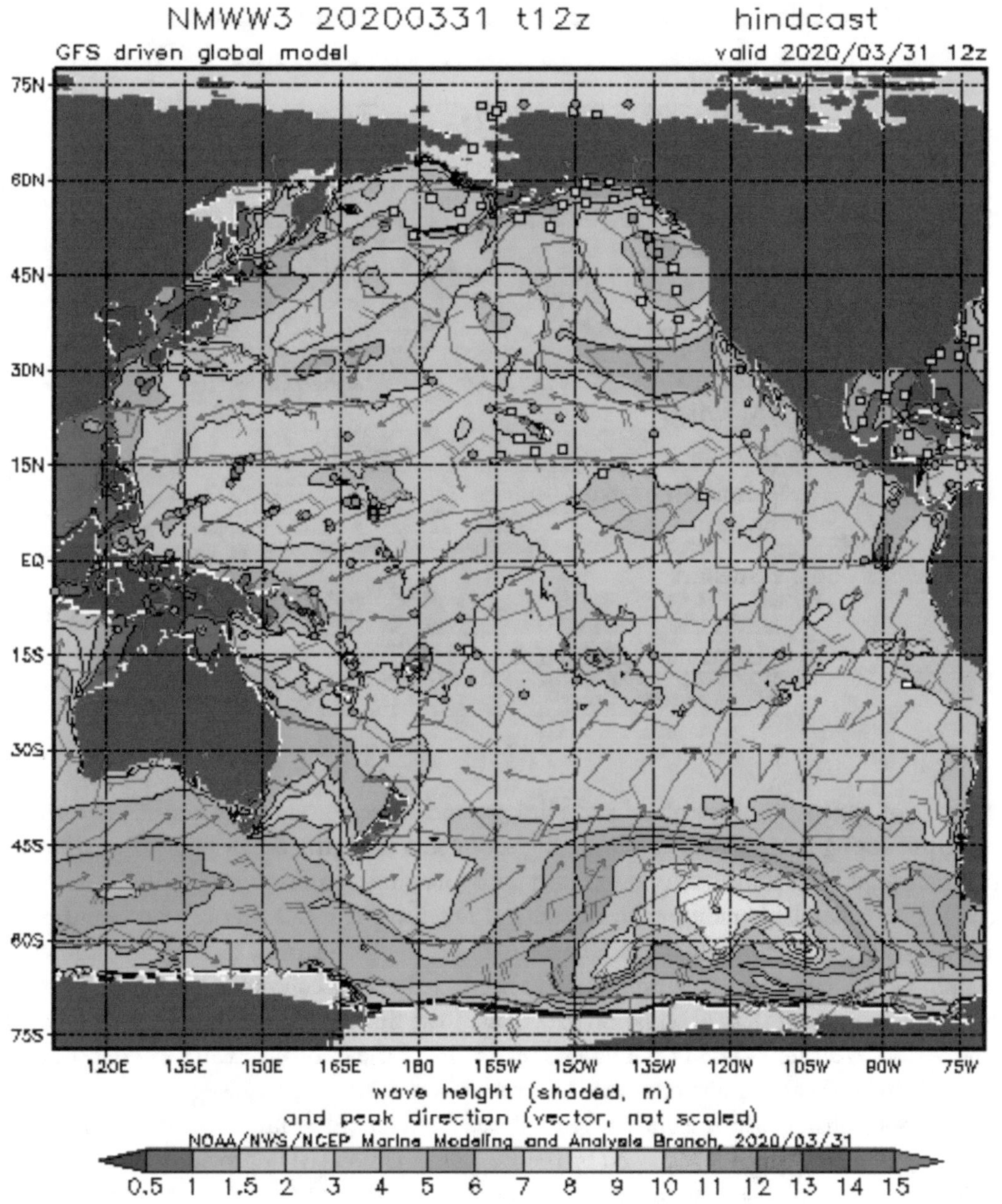

Figure 4.3: NOAA Multi-grid Wave Watch III (NMWW3) significant wave height forecasts for the Pacific Ocean Basin on March 31, 2020 (obtained from https://polar.ncep.noaa.gov/waves/). Significant wave height corresponds to mean wave height values that might be reported by an observer.

Table 4.3: General platforms used for the collection of environmental data

Platform
Aircraft
Argo floats
Autonomous underwater vehicle
Balloons
Commercial vessels
Data buoys
Drilling platforms
Kites
Moorings
Piers
Remotely operated vehicles
Research vessel
Satellites
Towers
Unmanned aerial vehicle
Unmanned surface vehicle

acteristics of water, the atmosphere and climate, and these vessels carry considerable equipment from coring devices used for collecting sediments from the ocean floor to weather balloons that send back information on atmospheric pressure, temperature, humidity, and wind speed via an expendable radiosonde. Additional detail on RV *GO Sars* can be found online at https://www.hi.no/en/hi/about-us/facilities/our-vessels/g.o.-sars. In recent years, unmanned autonomous vehicles and marine animals have been used as data collectors. Service Argos facilitates the collection of in situ observation data from floats that are used to support programs such as the World Weather Watch of the WMO, elements of the Global Ocean Observing System (GOOS), and animal tracking programs (see https://www.argos-system.org/).

Remote sensing instruments may be deployed from satellites, aircraft, or vessels. Programs such as Copernicus, which is managed by the European Union (EU), provide global data from satellites and ground-based, airborne and seaborne measurement systems. Important parameters inferred from radar altimeters aboard Sentinel satellites include sea surface height, sea ice extent, iceberg dimensions, sea surface winds, and significant wave heights. Copernicus Marine and Environmental Product Services have provided vital sea level altimetry products to validate ocean models. Many national hydrographic offices have installed microwave radar to fixed platforms such as piers to continuously measure water level fluctuations. Organizations such as

Table 4.4: Selected marine science instruments and samplers

Name	Function
Acoustic Doppler Current Profiler (ADCP)	Measures speed and direction for the water column.
Anemometer	Measures either total wind speed or the speed of one or more linear components of the wind vector.
Box corer	A marine geological sampling tool for collection of soft sediments in estuaries and the ocean.
Bongo Plankton Net	Fine mesh nets with a double-hoop that allow the collection of two samples in either horizontal, vertical, or oblique tows.
Conductivity, Temperature, Depth (CTD)	Determines the essential physical properties of sea water, i.e., salinity, temperature, and depth.
Gravity Corer	A weighted pipe that is allowed to free fall into the water to collect sediment within its hollow open pipe.
Magnetometer	Detects variations in the total magnetic field of the underlying seafloor.
Multiple Opening and Closing Net, with an Environmental Sensing System (MOCNESS)	A towed system of four individual plankton nets that are automatically opened and then closed.
Piston Corer	A long, heavy tube plunged into the seafloor to extract samples of mud sediment.
Rosette Sampler	A framework with 12–36 sampling bottles clustered around a central cylinder, where a CTD is often attached.
Sediment Trap	Containers that collect particles falling toward the sea floor.
Transmissometer	Measures the fraction of light, from a collimated light source, reaching a light detector a set distance away.
Van Veen Grab Sampler	A clam shell type of a bedload sampler.
Water Level Gages	Instruments that register the rise and fall of the water level.
Wave Buoy	Instruments that follow the movement of the ocean surface in order to measure parameters such as wave height, wave period, and wave direction.
For additional information on marine science instruments and samplers, see Nichols and Williams (2017).	

NOAA are implementing surface current radar to measure the speed and direction of ocean surface currents in near real time through Regional Associations that comprise the U.S. Integrated Ocean Observing System (IOOS) program. High-frequency radar systems provide important data which can be used during search and rescue. These radars which are sited along the shore can measure currents over a large region of the coastal ocean, from a few kilometers offshore up to about 200 km, and can operate under all weather conditions.

Ocean observing systems are designed to collect oceanic and atmospheric data and can thus be utilized in weather models to forecast ocean conditions in order to provide information to a variety of users including commercial and recreational mariners, emergency and coastal managers and responders, researchers and educators, and many more [Liu et al., 2015]. These networks may include data that is derived from drones, drifters, and numerical models. Many of the core measurements for the U.S. IOOS program are listed in Table 1.1.

Instrumented moorings are comprised of anchors, wires, chain, weights, floats, and buoys, which allow marine scientists and engineers to observe meteorological and oceanographic processes [Berteaux, 1976]. Instruments that may be fixed to a mooring to support oil drilling include conductivity and temperature probes for salinity and temperature and current meters. An attached surface buoy will likely support wave measuring sensors, cameras, and a weather station. Current meter data should be converted to speeds and directions. The recorded values which will be relative to magnetic north must be adjusted to geographic north by adding the magnetic variation. Speeds and directions should be checked for outliers or unexpected features such as spikes and noise bursts. Outliers may be identified by plotting scatter plots of the north-south and east-west directions of flow.

Before observations or data are ready for applications, they must be evaluated for errors and adjusted for calibration factors. Gaps in the data may be interpolated or techniques such as filtering may be applied to obtain values at standard times. Once data are processed they should be preserved in archives. This process may be documented in a data management plan. Nichols et al. [2017] delivered a data management plan to NOAA for the Southeastern Universities Research Association managed Coastal Ocean and Modeling Testbed project (see https://ioos.noaa.gov/project/comt/; Luettich et al. [2013]).

CHAPTER 5

Data Quality

Data quality refers to the condition of quantitative or qualitative variables that have been observed or calculated. Marine science and technology projects usually focus on quantitative data, i.e., numerical information that can be measured or counted. Qualitative data such as information from photographs, videos, and interviews provide evidence on the impacts of natural hazards such as earthquakes, floods, and tsunami or mishaps such as collisions, groundings, or structural failures. Data quality is especially important to develop reliable geospatial products from sources that may have different temporal and spatial scales. For the purposes of environmental characterization, data are generally considered high quality if real-world processes are correctly represented. In marine science and technology, data quality is a vital concern for professionals involved in activities such as scientific research, navigation, construction, weather forecasting, maritime security, emergency response, and forensic investigations. Incorrect data can originate from instrument errors, data transcription, incorrect algorithms, and poorly executed data migration and conversion projects. International organizations such as the IOC of UNESCO have developed standards that support projects that require nearly simultaneous observations over wide ocean areas [Dickson, 2010]. NOAA has defined data quality for ocean observing systems in a series of Quality Assurance of Real-Time Oceanographic Data (QARTOD) manuals (see https://ioos.noaa.gov/project/qartod/). NOAA (2005) reported findings from its second QARTOD workshop that was held from February 28 to March 2, 2005, on the calibration, metadata, and quality assurance/quality control needs of in situ currents, remote currents (high frequency radar), and waves.

5.1 DATA QUALITY ASSURANCE

Data quality assurance (QA) is the process of discovering inconsistencies and other anomalies in the data, as well as correcting errors by applying techniques that allow the removal of outliers or the filling of gaps. After a data set is collected, the data should be assessed to understand its quality challenges. Metadata, which are important to facilitate retrievability, will also identify QA procedures. Data should conform to standards that are published by IHO, IOC, WMO, NOAA, Federal Geographic Data Committee, etc. The WMO-IOC Joint Technical Commission for Oceanography and Marine Meteorology has published standards for oceanographic data (see http://www.oceandatastandards.net/). The output from QA includes measures of inconsistency, incompleteness, accuracy, precision, and missing data. QA ensures that quality control is being accomplished effectively. Researchers such as Baker [2016] and Waaijers and van der

Graaf [2011] explained how QA is essential to improving scientific rigor. Evidence-based management would rely on the QA of data that are used to characterize the environment to support management and decision making.

5.2 DATA QUALITY CONTROL

Data quality control (QC) is the process of controlling the usage of data for an application or a process. This process is performed both before and after a Data QA process, which consists of discovery of data inconsistencies and correction. The Data QC process uses statistical information from the QA process to decide whether or not to use the data for analysis or in an application. For example, if the Data QC process finds that an anemometer produces wind speeds with too many errors or inconsistencies, then it prevents that data from being used by harbor pilots which could cause disruption in shipping traffic. Providing invalid measurements from several sensors to vessels entering a seaport could cause a collision. Thus, establishing a QC process provides data usage protection. Operational organizations might implement real-time automated QC checks. Flagged data during QC may be further scrutinized to understand quality issues. Such analyses could involve inter-comparison with other types of data before flagged values are approved for release to users.

CHAPTER 6

Data Analysis

As the marine science and technology industry evolves, the volume of data that is available is dramatically increasing, and this implies that data analyses techniques are warranted to support effective, efficient, and accountable scientific or engineering projects. Many scientists and engineers use commercial software such as ArcGIS, Matlab, and Excel or open-source software such as Python or GNU Octave. Often times, scientists may obtain software licenses and maintenance for particular types of software. For example, scientists may utilize ENVI® software for the production of products that supports NOAA's Coastal Mapping and Airport Survey programs. Programs such as ENVI are particularly useful to extract meaningful information from imagery to make better decisions. These types of software systems all require the collection and archival of data for analysis. Many of today's data analysis programs can be deployed and accessed from the desktop, in the cloud, and on mobile devices, and can be customized through an application programming interface to meet specific project requirements.

6.1 TIME SERIES ANALYSIS

A time series is a set of observations such as water levels that are collected sequentially in time. These measurements are different than non-temporal data because data (e.g., water levels) are ordered before and after by some process (e.g., astronomical forces). The analysis of time-series data usually reveals significant, non-random relationships. This analysis considers the internal structure of the data such as autocorrelation, trends, and seasonal variation. Autocorrelation is a very general property of environmental variables and might be applied to detect recurrence or periodicity in a time series. Time series of sea surface temperature highlight a long-term trend of ocean heat uptake due to global warming. Polar scientists such as Noufal et al. [2017] have used temperature, salinity and current time series to identify seasonal variations in the Arctic Ocean. Ultimately, time series might be used to fit data to a model such as the autoregressive integrated moving-average (ARIMA) for forecasting purposes. Time-series analysis helps to identify the nature and strength of relationships and the prediction of future values. Tylkowski and Hojan [2019] used the ARIMA model to forecast sea level along the Polish Baltic coastal zone.

An important element in the analysis of observations from instruments such as water level recorders or current meters is the separation of tidal from non-tidal components of the signal [Pawlowicz et al., 2002]. For example, Nichols and Pietrafesa [1997] demonstrated the importance of evaluating the tidal, nontidal, and total signals to understand exaggerated ebb and flood tides through Oregon Inlet, North Carolina. The tidal and nontidal signals were removed using high and bandpass filtering techniques and then compared to time series of winds.

Meteorologists and oceanographers may make nowcasts or extrapolations for a very short-term period from other observations. Williams et al. [1993] demonstrated the nowcasting of currents using NOAA's first operational PORTS in Tampa Bay. These oceanographers collected 10 months of data to develop a linear regression between a permanent current meter located near the main channel of the Sunshine Skyway Bridge and a temporary current meter deployed 7.2 km away near a channel junction for Port Manatee. Once the temporary current meter was removed, the linear regression reported currents at the channel junction for users such as harbor pilots. NOAA's Center for Operational Oceanographic Products and Services (CO-OPS) operates and maintains the national network of PORTS in major U.S. harbors. PORTS is a decision support tool, which improves the safety and efficiency of maritime commerce and coastal resource management through the integration of real-time environmental observations, forecasts and other geospatial information. Nichols [1993] described the utility of the Tampa Bay PORTS following Hurricane Andrew.

6.2 SPECTRAL ANALYSIS

Environmental characterization of meteorological and oceanographic data requires applications such as interpolation, smoothing, filtering, and prediction. Interpolation and smoothing may be especially useful for observations that are taken at irregular locations and times. Filtering may be especially useful to separate a time series of water level fluctuations into its total, tidal, and nontidal components. Data may also be assimilated into a diagnostic or predictive model, either to help make forecasts or to enforce physical consistency constraints between multiple data types.

Many researchers such as Arabi et al. [2020], Dickey [1991], Janzen et al. [2019], and Xiong et al. [2018] have documented the need to integrate historical data, in situ data, imagery, and model output in order to capture the differing spatial and temporal resolutions that encompass phenomena that impact operations. This is especially true for the analysis of data from wave buoys and forecasting the wave field. During the biennial Ocean Waves Workshop[1] at University of New Orleans during 2019, Pieter Smit from Sofar Ocean Technologies described the Spotter buoy and methods to measure wind speed from a wave following buoy (see Fig. 6.1). Smit highlighted the use of spectral wave observations to estimate wind speed and direction [Voermans et al., 2020].

Spectral analysis is one of several statistical techniques necessary for characterizing and analyzing time series or sequenced data. Networked observing systems provide access to data that have been taken in one-, two-, or three-dimensional space, and time. Examples might be observations of fish near a platform, precipitation over an area, or significant wave heights. One important limitation is that the observations need to be consistent or equally spaced so that

[1]Ocean Waves Workshop retrieved from https://scholarworks.uno.edu/oceanwaves/.

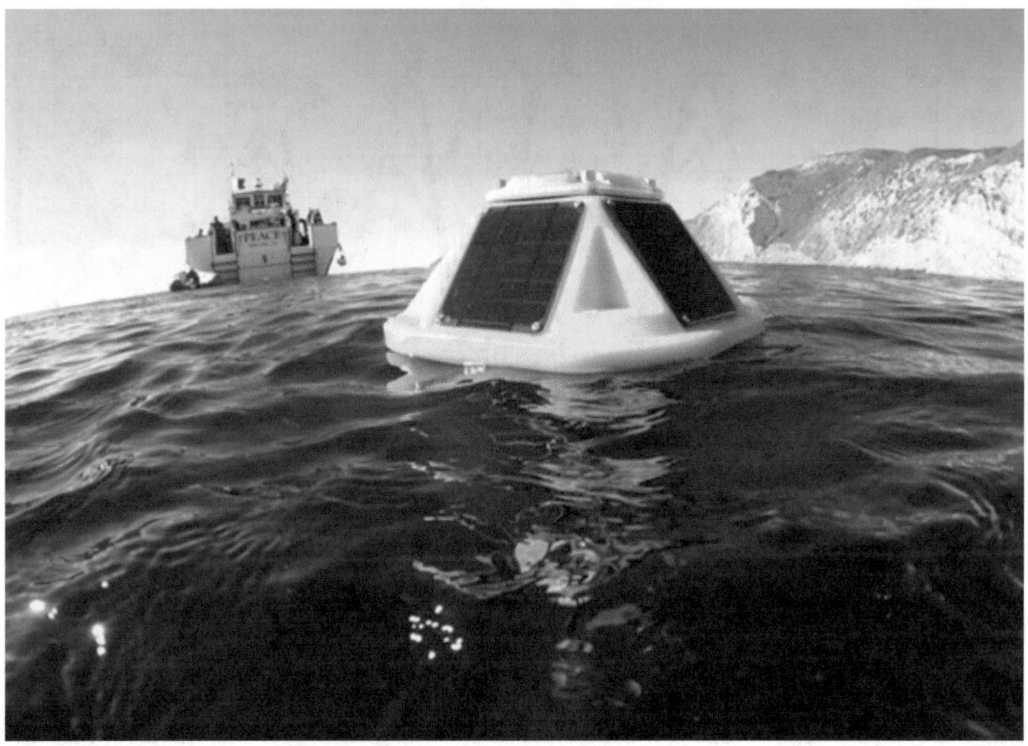

Figure 6.1: A drifting Spotter buoy was hand deployed from Dive Boat PEACE to measure wind, wave heights, wave periods, and wave directions along the southern California coast near Port Hueneme, CA (photo courtesy of Sofar Ocean).

oscillations of different lengths or scales may be identified. By this process, the observations in what is called the time domain are converted into the frequency domain. Data analysis is often easier in the frequency domain. For example, spectral methods have been shown to be very useful for nonlinear filtering. The observed cycles may suggest important factors that affect or produce such data.

The advent of modern computing has allowed for rapid and efficient computations of frequency spectra from measurements of time series. The Fourier series is a fundamental tool of harmonic analysis where an arbitrary periodic function can be expressed as an infinite sum of orthogonal sines and cosines. The expression of a periodic time series in terms of its sines and cosines allows for an examination of the relative contributions of its various harmonic components. When the underlying time series is discretely sampled, as are digitally sampled data, the Fourier components may be computed using the Discrete Fourier Transform (DFT). Highly efficient algorithms such as the Fast Fourier Transform (FFT) are widely available in most signal processing tools and packages such as Matlab or Python to compute the DFT. The FFT

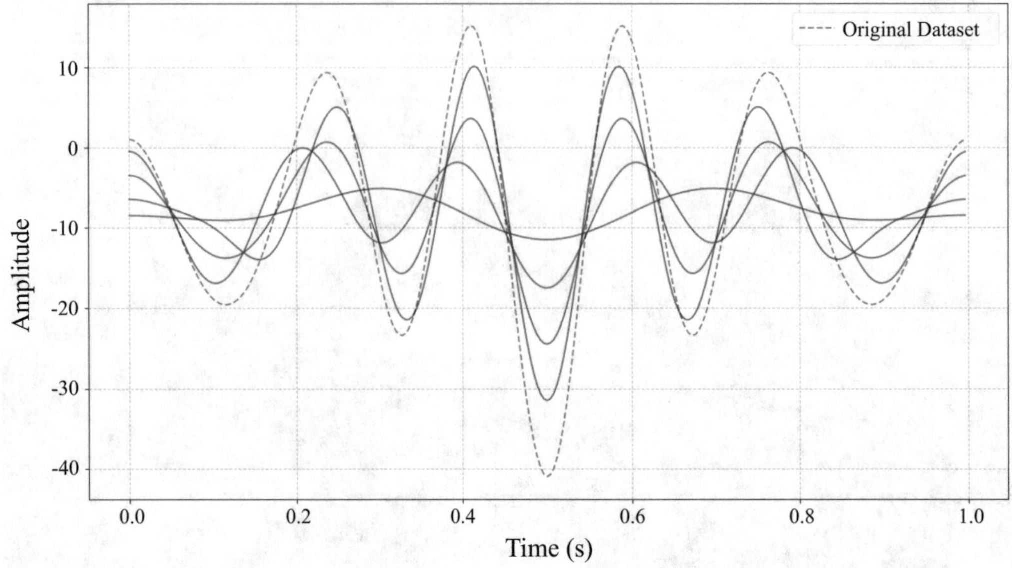

Figure 6.2: Four generalized Fourier components of a periodic signal.

forms the basic building block for more advanced spectral analysis such as the computation of power spectral densities. Figure 6.2 shows an example of the decomposition of an underlying periodic signal into its various Fourier components. As the number of Fourier components in the Fourier series is increased, the closer the match to original signal.

Measurements of waves are readily decomposed into their spectra. Typically, wave buoys measure horizontal and vertical displacements using either GPS or accelerometer-based methods [Raghukumar et al., 2019]. These data, when transformed into their frequency spectra, allow for an understanding of peak periods associated with measured waves (indicative of waves being dominated by swell seas or wind seas), and the direction that waves within a frequency band arrive from. Figures 6.3 and 6.4 show measured displacement time series on two wave buoys and their associated spectra. While the displacement measurements give a general idea of wave heights, much more information can be gleaned from an examination of the spectra, such as the various frequency components that contribute to wave displacements, and the energy associated with each component.

Researchers Bretschneider [1959], Hasselmann et al. [1973], Neumann and Pierson [1957], Ochi and Hubble [1976], and Portilla et al. [2009] described various types of spectral modeling approaches to estimate wave growth, a phenomenon providing valuable information for a variety of science and engineering applications. In general, developing seas tend to have a broader spectral peak while decaying seas have a narrower peak. While the JONSWAP Spectrum is used by many, the Bretschneider spectrum is sometimes used when the need for fully

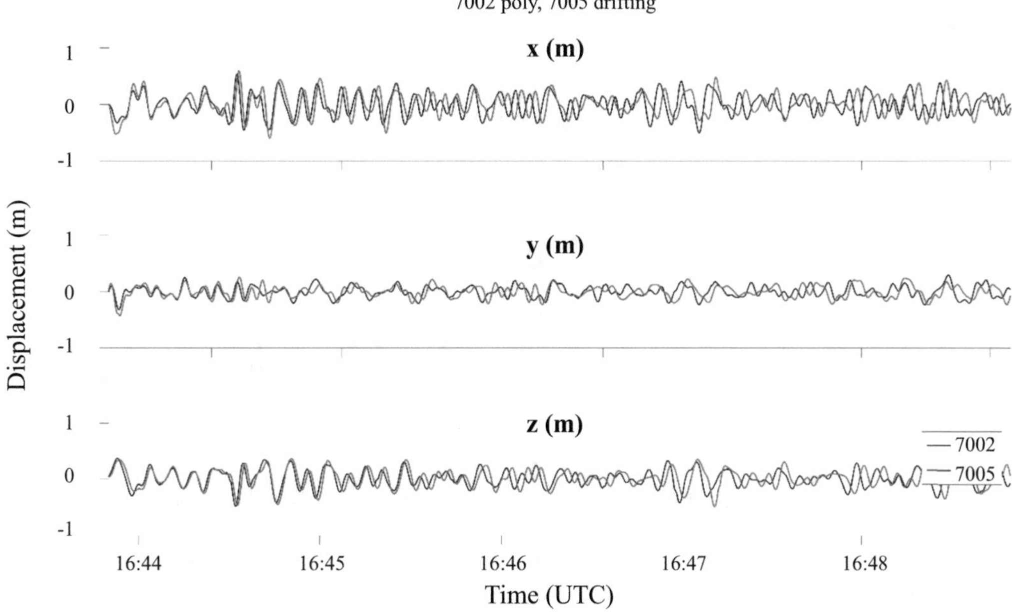

Figure 6.3: Raw displacement measured on a GPS wave buoy. From Raghukumar et al. [2019]. While wave-like motions can be seen, it is impossible to judge a primary wave period and wave direction.

developed seas is too restrictive. The Bretschneider spectrum may better represent seas under the influence of local winds with no swell energy present. For example, during the 2011 Ocean Waves Workshop at University New Orleans, representatives from Ocean Power Technologies, Inc. explained their use of the Bretschneider spectrum since it was more conservative for predicting power extraction [Edwards, 2011].

6.3 SPATIAL ANALYSIS

GIS is a powerful data-driven tool to visualize patterns in vector and raster data. Vector data are used to describe features on the earth that are represented by points, lines, or polygons. GPS instruments from mobile phones to survey equipment can be used to collect data for applications such as locating a tide station, representing the length of a pier, or marking the boundaries for a maritime forest, marsh, or coastal lagoon. Raster data are usually used to describe features on the earth that have been represented by pixels. Raster data may be derived from satellite imagery, aerial photography, digital elevation models, topographic maps, and numerical models, and may come in the form of a bitmap image such as a GeoTIFF or JPEG. GeoTIFF applies metadata standards which allow georeferencing information to be embedded within a Tagged Image File

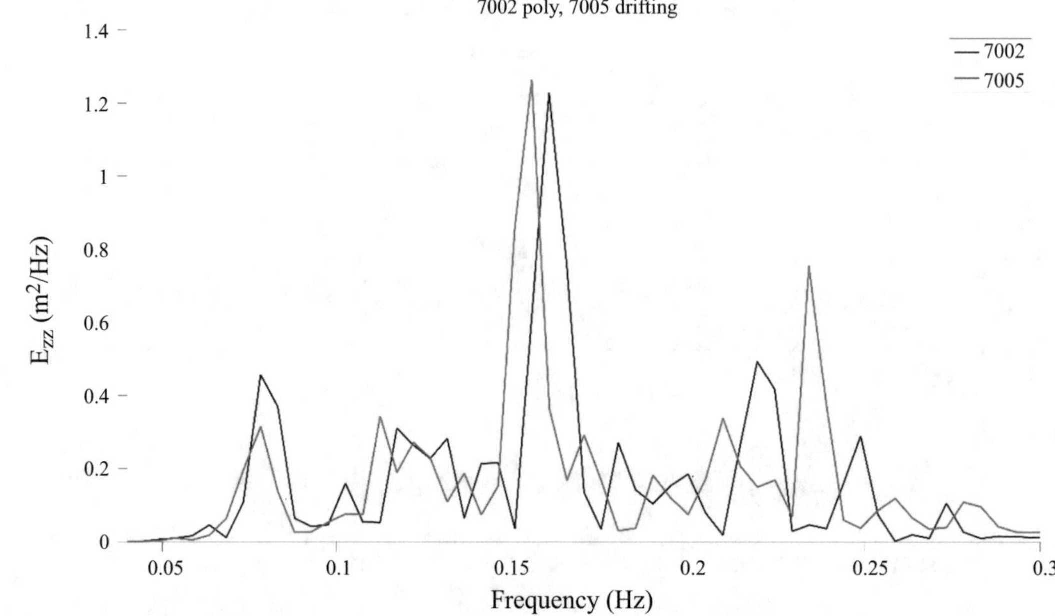

Figure 6.4: Power spectra computed from the displacement data in Fig. 6.3. The primary wave period and energy associated with this band are clearly seen.

Format (TIFF) file. Geospatial data are shared in a GIS through shape files, which consist of information regarding feature geometry, attribute data, spatial index to facilitate finding features, and projection.

Rules can be established to dictate appearance of features on a GIS map or numerical model output such as the use of colors to illustrate varying depths, temperatures, salinity, water quality, etc. The Gulf Stream System (GSS)provides a good example since its path undergoes meanders (time-varying lateral motions) that can shift this western boundary current tens of kilometers from its mean position along the east coast of the United States. Changes in the distance from the coast and the speed of the GSS have impacts on processes ranging from the nesting of sea turtles to sea level rise to search and rescue. Creating the rules for display of spatial data will depend on project requirements and of course the data, which may be derived from national archives, in situ sensors, imagery, and models. For example, the Rutgers University (RU) Center for Coastal Ocean Leadership (COOL) provides updated Gulf Stream products daily. Figure 6.5 depicts composite NOAA satellite imagery of the North Atlantic to highlight surface flow structure of the Gulf Stream on April 6, 2020. For the purposes of this discussion, this type of composite imagery displays the coldest waters as purple, with blue, green, yellow, orange, and red representing progressively warmer water. One can therefore distinguish between the cool, shelf water, which is blue in color from the core of the Gulf Stream which appears orange in color.

In addition to geo browsers such as Google Earth, various organizations around the globe provide archives that facilitate access to imagery-based products that can be applied to environmental characterization efforts. At NOAA, the Comprehensive Large Array-data Stewardship System is an example archive that serves NOAA and U.S. Department of Defense (DoD) Polar Operational Environmental Satellite (POES) data, NOAA's Geostationary Operational Environmental Satellite (GOES) data, and derived data. NOAA Satellite Maps, which can be accessed online, is a suite of interactive Earth-viewing tools that offer real-time, high-resolution satellite imagery from NOAA's most advanced geostationary and polar-orbiting satellites. Maps and geospatial products are also available from NCEI and GeoPlatform.org. The European Organization for the Exploitation of Meteorological Satellites or EUMETSAT provides a long-term archive of data and generated products, which can be ordered online from the EUMETSAT Data Center. Access to imagery and derived products from the EU's Copernicus program is made available through either the Data and Information Access Server or Conventional Data Hubs. University programs such as RU COOL routinely acquire satellite data from NOAA, NASA and EUMETSAT to support real-time ocean sampling by observatories such as the Mid-Atlantic Regional Association Coastal Ocean Observing System. These observations provide information for guidance and for weather forecasting. Example RU COOL products are available online at https://marine.rutgers.edu/cool/sat_data/.

Many marine science and coastal engineering problems will benefit from image classification, which could be considered as a specialized GIS application where remote sensing images are turned into meaningful map data, e.g., land use types or classes. The images are captured digitally and separated into different spectral bands including the visible spectrum (Red-Green-Blue) and others such as infrared [Bachmann et al., 2012a,b,c, 2018]. Based on the spectral information stored for each pixel, Bachmann et al. performed analyses known as classification to assign a pixel to a class such as land or water based on its similarity to reference pixels. This process helped in identifying the waterline or land-water boundary at time of imaging. Spectral information was also used to retrieve bathymetry [Bachmann et al., 2010a] and build bearing strength maps suitable for trafficability analysis [Bachmann et al., 2010b]. The process of "images to information" is essential in characterizing the environment [Eon et al., 2020] for decision makers who need to understand patterns and variability over time. Bachmann presented these capabilities at the 2019 NOAA Emerging Technologies Workshop[2] in Silver Spring, Maryland [Bachmann and Nichols, 2019].

Marine environmental characterization requires spatially explicit data at scales that are relevant to diverse applications or projects [Lecours, 2017]. For example, by combining satellite imagery with information gathered from benthic habitat surveys and geo-referenced photos and videos, one can effectively represent the environment and habitat types to facilitate communi-

[2]Emerging Technologies Workshop, Retrieved from https://nosc.noaa.gov/emerging-tech-workshop.php.

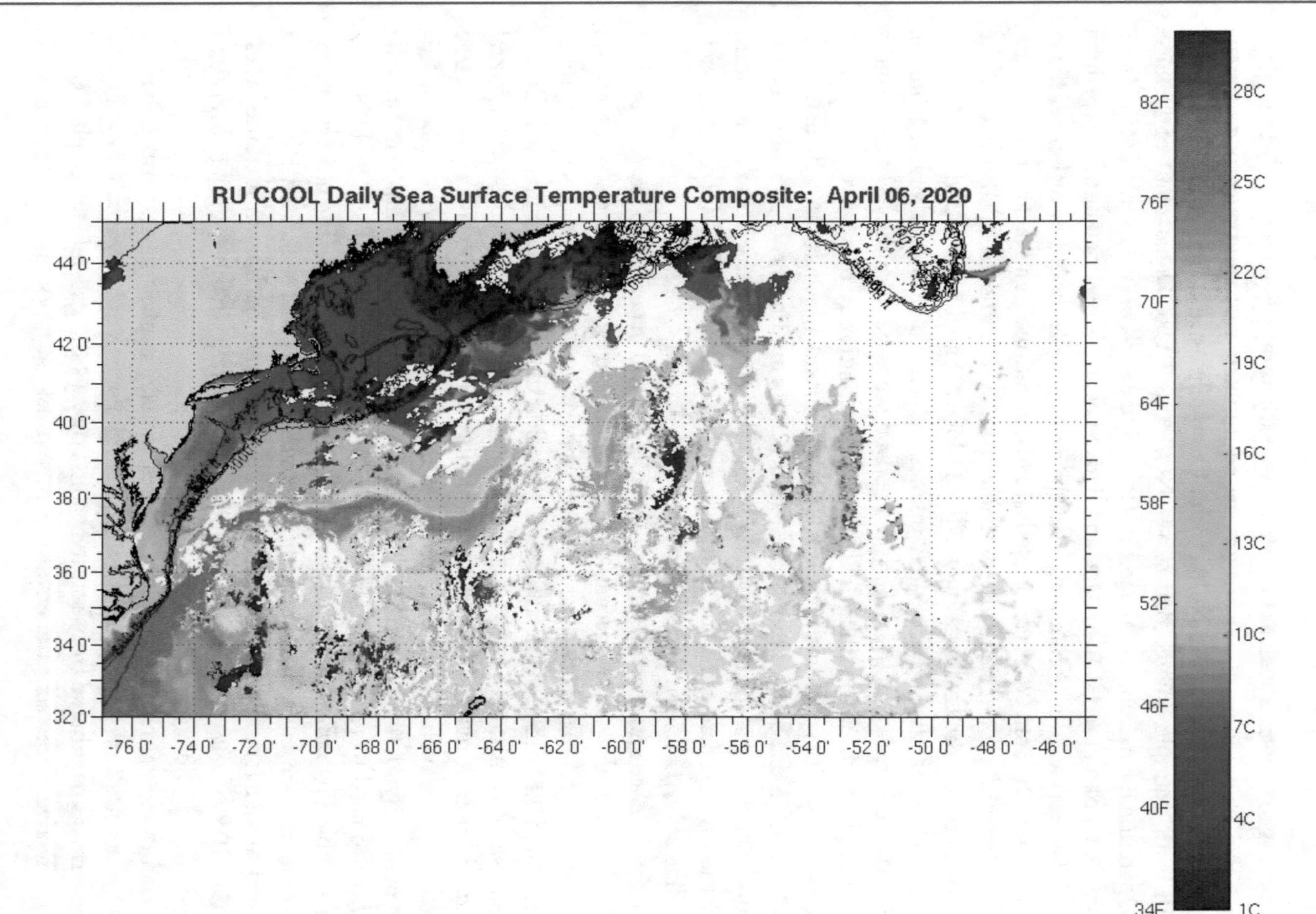

Figure 6.5: Image of the Gulf Stream based on sea surface temperature on April 6, 2020 (imagery obtained online from Rutgers COOL). The image illustrates meandering of the Gulf Stream downstream of Cape Hatteras, several warm-core rings shoreward of the Gulf Stream, and cold-core rings seaward of the Gulfstream.

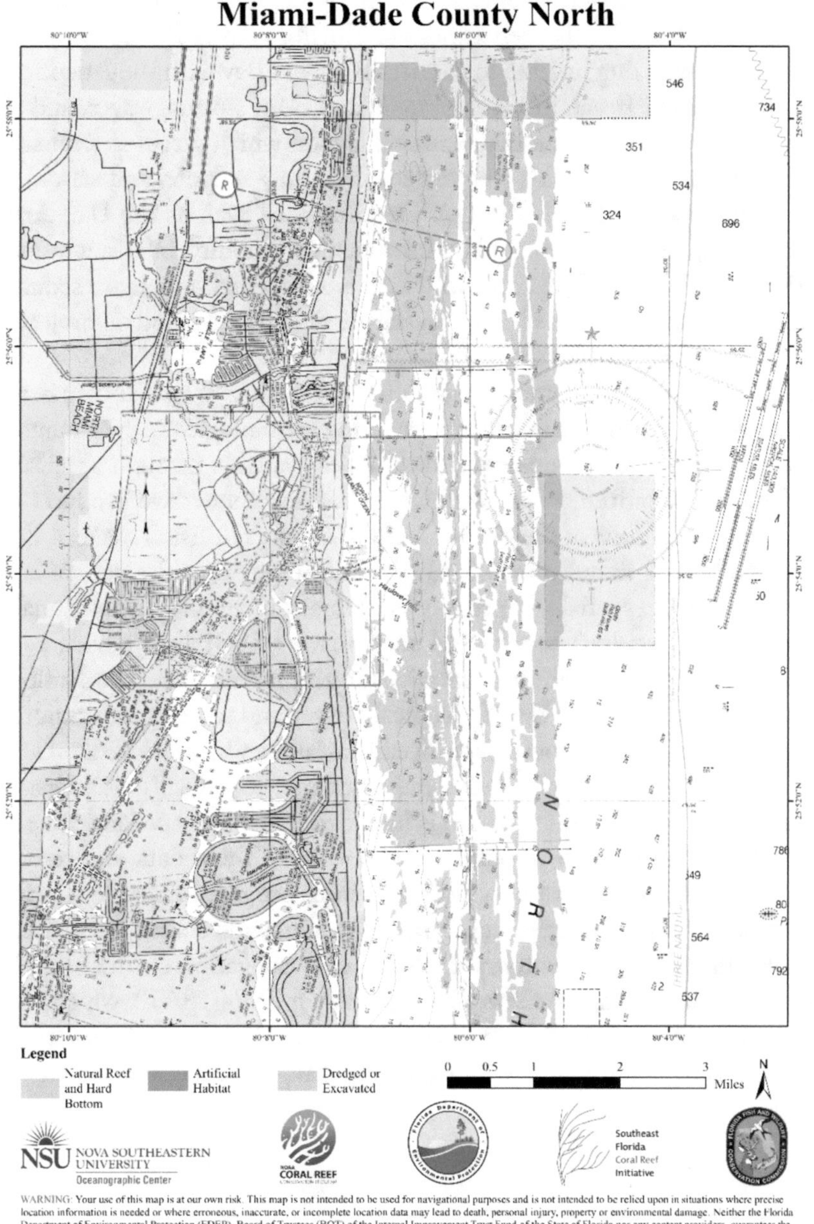

Figure 6.6: Example habitat map which can be used to support coastal engineering and beach management activities.

cation among different stakeholders concerned with topics such as coastal resilience. Figure 6.6 is a habitat map from the Florida Department of Environmental Protection covering Bakers Haulover Inlet, which was dug in 1925 to connect Biscayne Bay with the Atlantic Ocean. Barrier islands such as Miami Beach protect the mainland from storms, waves, and sea level rise. Miami Beach also protects Biscayne Bay, a coastal lagoon which serves as a nursery for a variety of commercially important finfish (spotted seatrout, gray snapper, and snook) and shellfish (spiny lobster, stone crab, and pink shrimp). Organizations such as the U.S. Army Corps of Engineers provide support in understanding inlet channel stability, inlet migration, shore erosion, and activities such as beach restoration which are designed to balance sediment budgets. Sediment needs for locations such as Miami Beach are established based on project plans, storm frequency, construction losses, and sea level change.

Big data include the ever-increasing quantity and quality of spatial data becoming available to characterize the marine environment, emerging technologies such as unmanned surface vehicles to collect these data and analytical techniques available through applications such as GIS. Researchers and practitioners are working to develop innovative products to better apply spatial data representations of the environment to support decision makers. For geoscience information in the United States, the U.S. Geological Survey provides geospatial data, where information is presented in spatial and geographic formats, including The National Map, Earth Explorer, GloVIS, and LandsatLook. Helsel and Hirsch [2002] also provide a detailed textbook that discusses the analysis and presentation of geographic data. The National Geologic Map Database is supported by the U.S. Geological Survey and serves as an authoritative source for geoscience information. To provide improved coastal data in Florida, the U.S. Geological Survey and the Florida Institute of Oceanography embarked on a Florida Coastal Mapping Program [Hapke et al., 2019]. This program is intended to facilitate collaborative research that provides improved access to high-resolution seafloor data. These data and resources such as Florida's Historic Shoreline Database support numerous projects related to Florida's extensive coastal zone including beach management, habitat mapping, and coastal resilience. Derived information on shoaling rates and hydrographic surveys of navigation channels can be used with the U.S. Army Corps of Engineers' *Corps Shoaling Analysis Tool* (CSAT), which identifies likely shoaling locations within a navigation channel [Dunkin et al., 2018]. Tools such as CSAT are important to objectively determine how shoaling has changed in the wake of extreme weather.

CHAPTER 7

Key Challenges

Engineering projects need to consider environmental factors in activities that generally comprise the requirements engineering process—elicitation, analysis, specification, validation, and management. The requirements cycle helps to ensure that a system can operate reliably while remaining within the range of its operating specifications. It is also important that environmental scientists are aware of engineering needs. In this way, the most useful historical data can be mined, optimally sited sensors can be installed for monitoring, appropriate models can be set-up, and variations such as seasonal changes can be identified. Further, through collaboration among engineers and scientists, critical products can be designed to support sustainability and resilience. Signals and alarms can also be implemented to notify operators that conditions are outside of the systems' established operating specifications.

Owing to the vast scale of the oceans and the difficulty and expense in making measurements, data gaps may exist that will challenge project planning. Since the ocean is undersampled, one might apply numerical models to simulate environmental conditions. However, if the modeler runs the model with incorrect initial conditions, the user may be provided with incorrect information on environmental loads. A lack of observational data in the ocean can result in operationally unaceptable uncertainty in model output. The oceanographic data shortfall is being addressed by the science community through the installation of ocean observing systems, worldwide. Today's GOOS is a global system for sustained observations of the ocean and is administrated by the IOC (see https://www.goosocean.org/).

Engineering projects must capitalize on science and technology innovations that address issues related to data sparsity. For example, unmanned systems are making ocean observations more available. Defense organizations have already demonstrated the utility of unmanned systems for data collection to support mapping, mine clearing, underwater construction, and surveillance. These unmanned systems are integrated with state-of-the-art sensors which are smaller in size, weight, and consume less energy. Many government funded projects also require data sharing which has contributed to the development of metadata and standards to facilitate the exchange of oceanographic data and information. One of the principles behind the development of the GOOS is open exchange of data and metadata with minimum possible cost, delay, and restriction.

Collecting optimal data on environmental factors will be essential to describing relationships between variables that impact engineering design and system operation. An ongoing area of research that has improved the skill of forecasts has involved data assimilation, where observational and model information are combined to provide an estimate of the most likely state of the ocean and its uncertainty. Modeling will be key to address specific information requirements at varying temporal and spatial scales. Optimal observations, including historical information and data extracted from imagery, can be combined with model output to fill information gaps. They have the largest impact on generating reliable model solutions and information that supports decision makers.

The general lack of oceanographic information can lead to uncertainty in decision making and outcomes analysis. Reliable information is necessary to implement and evaluate problems effectively, especially challenges related to sustainability and resilience. By collecting environmental data before, during, and after a project, one can ensure that the project does not compromise the environment and is able to cope with climate change phenomena such as sea level rise. Sustainable engineering requires an interdisciplinary approach in all aspects of engineering, but especially the environment. Project managers must determine what data to collect, how to collect, when to collect, and what tools to use to synthesize the data. Coping with climate change requires the integration of environmental, engineering, and community resilience. Toward this, public and private organizations are working together to identify needs and develop the suite of resources (e.g., data, information, tools, training) that are needed to cope with issues such as recurrent flooding, erosion, and accelerated shoreline retreat. Successful approaches that have been endorsed by organizations such as NOAA, U.S. Army Corps of Engineers, and NGOs such as the Nature Conservancy involve assessing hazard risks and community vulnerability, identifying nature-based solutions, taking conservation and restoration actions, and measuring the effectiveness of actions that are intended to reduce flood risk.

CHAPTER 8

Conclusions

Scientists and engineers benefit from a solid foundation in the geosciences. Geoscience is the study of the Earth—its oceans, atmosphere, rivers and lakes, ice sheets and glaciers, soils, its complex surface, rocky interior, and metallic core. Environmental data are necessary to characterize the environment and to provide environmental information to support engineering projects. Data describing wind, wave, current, and other external forces are essential to compute marine environmental loads, which can impact structures such as platforms, bridges, and towers as well as ride quality and seakeeping for vessels.

Marine environmental monitoring and characterization provides a foundation for a variety of operations from docking and undocking vessels to marine construction to oil spill response. Data collection and communication methods to provide relevant physical, chemical, and biological information that impacts operations are critical for supporting maritime operations by reducing risks and ultimately protecting property and saving lives. Measuring systems such as Canada's SmartATLANTIC provide valuable support to mariners by providing accurate and real-time meteorological/hydrological data to include high-resolution forecasts of weather and sea conditions. Most of the data from this network are archived and made available via online data request forms. Nested regional modeling approaches may provide spatially extensive information. For example, Australia's eReefs modeling system also helps to understand issues surrounding coral bleaching and may be used to mitigate the impact of incidents, reduce the risk of an oil spill, and support the region's ship-based trade. Figure 8.1 illustrates a combination of temperature, wind, salinity, and current, key hydrodynamic parameters impacting the Great Barrier Reef off the coast of Queensland, Australia.

GIS provides a valuable tool for spatial analysis. Projects may benefit from the design and population of a file geodatabase, which is a collection of files in a folder on disk that can store, query, and manage the spatial and nonspatial data for a particular project [Bachmann et al., 2009b, 2012a,b,c, 2016]. Information in the geodatabase includes historical information such as time series, project data from instruments, photo transect, video transects, overhead imagery, and model output. These data may also be used as ground truth for the calibration and validation of imagery and model output. GIS-based marine environmental characterization provides a way to manipulate big data to produce objective information for projects. Example products that support coastal construction and maritime operations are provided in Table 8.1.

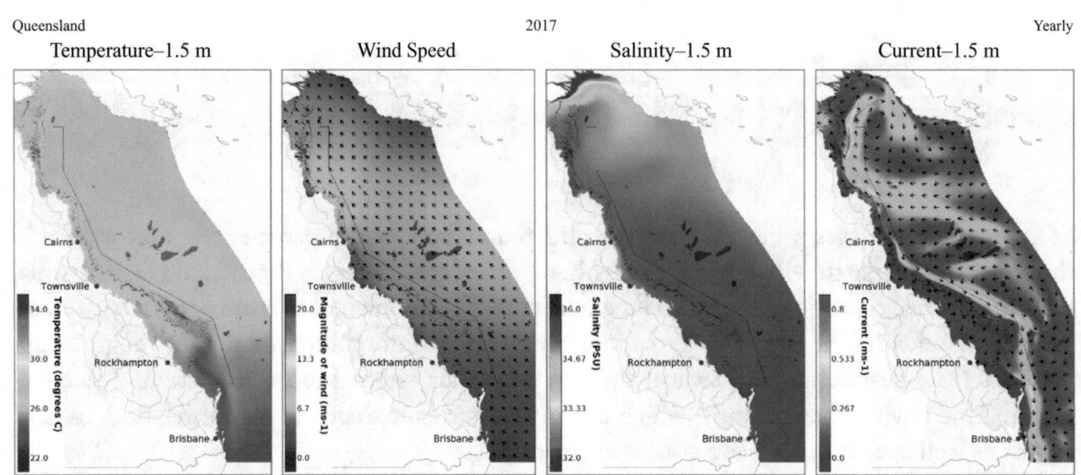

Figure 8.1: Averaged environmental conditions common to the Great Barrier Reef during 2017 from the eReefs Hydrodynamic model (Aims eReefs Visualization Portal, retrieved from https://aims.ereefs.org.au/aims-ereefs).

Table 8.1: Spatial analysis is dependent on various type of observations. Examples include water level data, depth data, profiles of salinity, temperature, fluorescence, oxygen, suspended sediment, current profiles, and wave data. These data are used to create a variety of traditional products.

Product	Example Application
Bathymetry	Safe navigation
Chlorophyll concentration	Harmful algal bloom mapping
Discharge	Fate of dredge material
Flow distribution and turbulence	Dissipation of a marine spill
Oxygen depletion	Mapping extent of hypoxia or dead zones
Tidal constituents	Flood forecasting
Water masses, stratification, mixing	Piloting an unmanned underwater vehicle
Wave height, period, and direction	Seakeeping and ride quality
Observations made directly or parameters which are calculated as function of other data are associated with uncertainties.	

Glossary of Environmental Terminology

Engineering projects throughout history have involved teamwork. Modern projects may include variously sized organizations or groups of individuals that are geographically dispersed. Project challenges can be addressed by combining expertise from multiple subdisciplines, especially environmental sciences (e.g., geology, meteorology, oceanography, remote sensing).

The multidisciplinary nature of projects that are impacted by environmental factors creates the need for a glossary to define foundational terms. Definitions for these terms and others may be found in references such as *Tomorrow's Coasts: Complex and Impermanent* [Wright and Nichols, 2019]; American Meteorological Society *Glossary of Meteorology*, 2nd ed.; *Encyclopedia of Marine Science*, 2nd ed., [Nichols and Williams, 2017]; *Tide and Current Glossary* (CO-OPS, 2000); and the *Coastal Engineering Manual* [U.S. Army Corps of Engineers, 2002]. Some of the following definitions have been modified from the original, and the reader is urged to consult a textbook if greater detail is required.

Abiotic factors: Non-living chemical and physical components of the environment that affect living organisms and ecosystems. Examples of abiotic factors include currents, grain sizes, nutrients, salinity, soil, sunlight, temperature, and water levels. Pietrafesa and Janowitz [1988] described physical oceanographic processes affecting the transport of larval fish through coastal inlets.

Accretion: The accumulation of land by natural causes such as wind, waves, river discharge, and longshore currents. There is also an increased potential for accretion when wave energy impacting a shoreline is reduced. In tidal marshes, grasses such as smooth cordgrass (*Spartina alterniflora*) trap sediments and create peat, they may rise with sea level in a process that may be called "wetland accretion." Boyd and Sommerfield [2016] discussed biotic and abotic factors causing accretion at study sites located in the Prime Hook National Wildlife Refuge on the southwestern shore of the Delaware Bay. They found that marsh lands were accreting at similar rates to the local relative rates of sea level rise.

Aerosol: Liquid or solid particles that are dispersed in a gaseous system such as air (e.g., dust, fumes, mists, sea spray, smoke, and fog). Xiao et al. [2018] described aerosol particulates collected in the western north Pacific from China's Research Vessel *Kexue*.

Anthropogenic: Caused or produced by humans as opposed to natural processes. Air and water pollution occur when harmful substances including particulates and biological molecules are introduced into Earth's atmosphere and hydrosphere. These releases may cause diseases, allergies or death of humans; it may also cause harm to other living organisms such as animals and food crops, may damage the natural or built environment, and contribute to climatic changes. Wright and Nichols [2019] described anthropogenic and natural coastal changes and suggest the integration of environmental, engineering, and community resilience efforts.

Aquifer: An underground layer of permeable rock (e.g., sand, gravel, and sandstone) that acts as a reservoir for groundwater. Aquifers fill with water that drains into the ground. Wells drilled into aquifers provide water for drinking, agriculture, and industrial uses. Aquifers can be depleted or dry up when people drain them faster than nature can refill them. When too much water is removed from the soil, the soil may collapse, compact, and drop contributing to land subsidence. Saltwater intrusion into freshwater aquifers can lead to groundwater quality degradation. Bloetscher et al. [2014] described more than 200 United States aquifer storage and recovery projects, a water resources management technique where water is stored underground during wet periods for recovery during dry periods.

Atmosphere: The gaseous envelope surrounding the Earth. The atmosphere consists almost entirely of nitrogen (78.1% volume mixing ratio) and oxygen (20.9% volume mixing ratio), together with a number of trace gases, such as argon (0.93% volume mixing ratio), helium, greenhouse gases such as carbon dioxide (0.035% volume mixing ratio), and ozone. The atmosphere also contains water vapor, clouds, and aerosols. Emissions of sulfur dioxide and nitrogen oxide, which react with the atmospheric water molecules, contributes to phenomena such as acid rain. Mueller et al. [2011] discussed the impact of exhausts of particulates from seaports and populated coastal areas that are associated with heavy ship traffic.

Atmospheric pressure: The pressure exerted by the earth's atmosphere; it decreases with altitude above sea level. At sea level, the standard atmospheric pressure is equal to 1013.25 mbar (14.7 psi). Atmospheric pressure is an important parameter which determines wind and weather patterns. Researchers such as Gillett et al. [2003] have shown that average sea-level air pressure has risen in recent years over parts of the subtropical North Atlantic Ocean, southern Europe and North Africa while dropping in locations such as the poles and the North Pacific Ocean. Increases in sea-level pressure decrease water level while decreases in pressure increase water level.

Bar: A submerged or emerged embankment of sand, gravel, or other unconsolidated material built on the seafloor in shallow water by waves and currents. Storms tend to carry sand seaward, forming offshore bars and then this sand migrates landward during calm weather. A bar is also a metric unit of pressure equal to the average pressure of the atmosphere at sea level or 106 dyne cm^2 (106 barye), 1000 mbar, 29.53 in. of mercury. Within the ocean water column, the pressure increases by one bar with every 10 m (33 ft) of water depth.

Barrier island: A long, usually thin, sandy stretch of land, oriented parallel to the mainland coast between two inlets. They face the ocean on one side and a lagoon on the other side. Barrier islands provide a natural line of defense against storms that threaten coastal communities by absorbing wind and wave energy. Pilkey et al. [1998] described North Carolina's barrier island system that is known as the Outer Banks.

Beach: The location along a shoreline where the sediment is in motion, being moved by winds, waves, tides, and currents. Loose unconsolidated sediments such as sand that are transported to places suitable for deposition will form a beach. They are dynamic and vary in length, width, composition, and permanence. Sea level rise will affect beaches differently because of processes such as isostatic or postglacial rebound where land masses that were once depressed by ice sheets rise while those that were raised at the edges of the glaciers are now sinking. Beaches will evolve in responses to forces such as sea level rise and coastal development. In many countries beaches may be closed due to the presence of waterborne pathogens in the ocean. The U.S. EPA provides online information on the U.S. beaches at https://watersgeo.epa.gov/beacon2/.

Beach material: Granular sediments, usually sand or shingle, moved by waves and currents. Dams built for flood control and water catchment along the rivers leading to the coast inhibit the transport of sediments. The lack of sediment from rivers accelerates coastal erosion.

Beach nourishment: The restoration of a beach by placing beach material directly onto the beach. Beach nourishment is expensive and only a temporary solution to counter sediment starvation along severely eroding coasts. During 2017, the towns of Duck, Kitty Hawk, Kill Devil Hills, and Southern Shores along the Outer Banks of North Carolina nourished miles of beach by pumping sand from the ocean floor onto the beach to build up eroded areas. These efforts are intended to create a wider beach area to protect and enhance the shoreline. Bird and Lewis [2015] described beach dynamics and highlighted issues related to shore protection techniques such as beach renourishment.

Benthic habitat: Physically distinct areas of the seafloor associated with suites of species that consistently occur together. Species of animals and plants that live on or in the bottom are known as the benthos. The creation of *benthic habitat* maps is essential to the management and analysis of these important ecological resources. NOAA examples may be accessed at https://products.coastalscience.noaa.gov/collections/benthic/default.aspx.

Bight: A geographic term describing a bend or curve in the shoreline such as a large and receding bay. Examples include Bight of Benin, Canterbury Bight, Flemish Bight, Great Australian Bight, Mid-Atlantic Bight, New York Bight, North and South Taranaki Bights, and Southern California Bight. James et al. [2001] described the temperate carbonate platform that forms the shelf of the Great Australian Bight.

Biomass: Biological materials including organic material (both living and dead) from above and below, e.g., mangrove trees, marsh grasses, submerged aquatic vegetation, roots, and animals and animal waste. Walters and Kirwin [2016] discussed the resilience of marshes to sea level rise at the Virginia Coast Reserve, a Long-Term Ecological Research site near the mouth of the Chesapeake Bay (see https://lternet.edu/).

Breaker: A wave breaking on the shore. As an ocean swell propagates from deep water to shallow water, it will undergo transformations through the effects of refraction, diffraction, and/or shoaling until the wave becomes unstable. Once the wave reaches a critical height, it will overturn on itself and break. Based on the dissipation of energy owing to factors such as beach slope, wind conditions, and nearshore currents, the breaking will result in collapsing, plunging, spilling, and surging surf. These breakers and the ensuing longshore currents tend to pick up and move sand up and down the shoreline, generally offshore during winter storms and back toward land during the summer's fair weather. Feddersen and Trowbridge [2005] discussed a wave breaking and longshore current modeling based on data collected at the U.S. Army's Field Research Facility during the Duck94 field experiment.

Cape: A prominent land area jutting seaward from a continent such as Cape Comorin in Asia or a large island such as Cape Engaño in the Philippines. A cape usually represents a marked change in the trend of the coastline. Capes can be formed by glaciers, volcanoes, and changes in sea level and have a relatively short geologic lifespan owing to natural erosion by winds, waves, tides, and currents. Some other cape examples include Cape Cod in Massachusetts and Cape Hatteras in North Carolina and Cape of Good Hope in South Africa.

Climate: The prevailing weather conditions for a region. Climatic elements include temperature, air pressure, humidity, precipitation, sunshine, cloudiness, and winds, throughout the year, averaged over a series of decades. Climate differs from weather, in that weather only describes the short-term conditions of these variables in a given region. Example statistics may include labels such as normals, means, and extremes. Normals are generally based on the distribution of all observations over a long-period such as 30 years (e.g., 1981–2010). Means will refer to the average of the maximum and minimum temperatures over a particular period such as a day. These means do not provide any information about how the observations are scattered around the mean, i.e., whether they are tightly grouped or broadly scattered. The extremes are usually observations lying in the most unusual ten percent. Climatic changes take hundreds, thousands, even millions of years to change while the weather can change in just a few hours. Climate change, therefore, is a change in the typical or average weather of a region or city. Earth's climate is always changing as evidenced by climatic records that indicate warmer and cooler periods, each lasting thousands of years. Trends from quality-controlled observations currently indicate that the Earth's climate has been warming. The Earth's warming climate is linked to changing rainfall patterns, decreasing snow and ice cover, and rising sea levels. It should also be noted that climatic data can be confounded by various non-climatic effects, such as the relocation of weather stations, land-use changes, changes in instruments, and observational hours.

Coast: A strip of land of indefinite width that extends from the shoreline inland to the first major change in terrain features. In this book, we use the term "coast" to imply the entire region influenced by land-sea interaction, typically from the shelf break to about 100 km inland. Coasts are sensitive to sea level rise, changes in the frequency and intensity of storms, increases in precipitation, and water temperatures. Severe storms generate surge, waves, and currents that can move large amounts of sediment, destroy roads, buildings and other critical infrastructure as well as alter natural habitats.

Coastal or sea cliff: A very high and nearly vertical land feature forming near the ocean owing to processes such as weathering and erosion. The upper cliff is transformed by weathering while erosion wears away the cliff base. Erosion rates vary depending on sea level rise, wave energy, cliff slope, beach width and height, and rock strength. In some locations, cliff retreat rates have accelerated owing to sea level rise. Limber et al. [2018] analyzed multiple models to describe 21st-century cliff retreat rates.

Coastal inlet: A passage separating two barrier islands and connecting a lagoon or bay with the ocean. These connections are important for sediment transport and fish migrations. Changes in the morphology and behavior of an inlet and its associated sediment bodies can have far-reaching impacts on adjacent coastlines. Kraus [2007] described approximately 16 Texas inlets that connect inland coastal waterways and rivers to the Gulf of Mexico and processes that control their creation, stability, and navigability.

Coastal zone: The dynamic region where land meets water. These regions have become increasingly important because they support large populations and continually change owing to natural (tropical cyclones and tsunami) and man-made (coastal structures) phenomena. Climate change can affect coastal zones through shoreline erosion, coastal flooding, and water quality. Coastal zone management efforts ensure the health and stability of the coast, both environmentally and economically, into the long-term future.

Coastline: The land and water interface, which is also called the shoreline. The coastline is constantly eroding and forming headlands, bays, and cliffs. The weaker or softer rock, such as sandstone, is eroded fastest leaving more resistant rock types, such as granite, which might remain as a sea cave, arch, or stack. Remote sensing surveys of the coast call the land and water interface at time of imaging the waterline. Waterlines can be used to map shoreline changes following severe storms.

Compound flooding: Severe flooding related to the simultaneous additive effects of multiple causes, for example river (fluvial) flooding, pluvial flooding by torrential rainfall, and storm surge. Wahl et al. [2015] described flooding in low-lying coastal regions owing to the combined impact of storm surge and heavy precipitation.

Cyclone: A system of winds that rotate about a low-pressure center. Rotation is clockwise in the Southern Hemisphere and anti-clockwise in the Northern Hemisphere. In the Indian Ocean, the term cyclone refers to a tropical cyclone or what would be called a hurricane in the Atlantic Ocean. An example storm was Cyclone Nargis which developed on April 27, 2008 in the Bay of Bengal where it moved eastward and intensified to attain peak winds of approximately 215 km/hr (130 mph) before making landfall in the delta region of the Ayeyarwady River (Irrawaddy River) of Myanmar on May 2, 2008, and finally dissipating near the border of Myanmar and Thailand. This cyclone caused catastrophic destruction in Myanmar with a death toll that was estimated to exceed 135,000 people.

Dead zone: A low-oxygen (hypoxic) area in the world's oceans and lakes that is caused by excessive nutrients (e.g., nitrogen and phosphorus) or eutrophication. The enrichment of the water body with nutrients induces growth of phytoplankton and algae and due to the biomass load results in oxygen depletion. The resulting hypoxic areas usually appear near the bottom where the oxygen concentration is so low that benthic and demersal animals suffocate as evidenced by associated fish kills, distressed animals, and sulfur-oxidizing bacteria that cause the anaerobic sediments to turn black. While natural phenomena can cause hypoxia, the improper use of fertilizers and inadequately treated or untreated sewage exacerbates the problem as water high in nutrients drains into streams, rivers, and the ocean. Dead zones have been observed in the northern Gulf of Mexico, North Sea, Black Sea, northern Adriatic, Chesapeake Bay, Kattegat Strait, and more.

Delta: A seaward prograding and fan-shaped sediment body deposited at the mouth of a river. A bird-foot delta (e.g., the Mississippi River delta with its levee-bordered distributaries) extends seaward and resembles the claws of a bird. A cuspate delta such as the Ebro delta in Spain extends into the sea like an arrow. An arcuate delta forms owing to tides and waves as in the case of the rounded convex outer margin of the Nile delta in Egypt. Deltas provide fertile lands, support commercial fishing, and tend to be locations with oil and gas deposits, but these low-lying areas are also associated with severe flooding. For additional information, see the World Delta Database (Coleman and Huh [2004]; www.geol.lsu.edu/WDD).

Descriptive statistics: Observations are described by indexes such as means and standard deviation. By building long-term databases of observations, an analyst may evaluate changes from the baseline.

El Niño: Coupled ocean-atmospheric interactions across a broad expanse of the equatorial Pacific Ocean with a global impact on weather patterns. The condition is marked by a significant increase in sea surface temperature over the eastern and central equatorial Pacific that occurs at irregular intervals, generally ranging between two and seven years. Warming of the Humboldt current and shifting wind patterns as a consequence of El Niño were determined to also impact the upwelling of nutrient rich waters (off the coast of Peru and Chile) for marine species such as anchoveta (*Engraulis ringens*), which in turn impacts other populations and makes the ecosystem very vulnerable to conventional intensive fishery practices. Ward et al. [2014] link the impacts of El Niño Southern Oscillation caused flooding to economic damage.

Erosion: The movement of weathered or decomposed rock material or soil by natural forces such as winds, waves, and currents. Coastal erosion may also be caused by river damming and diversion, owing to the loss of sediment supply to the coast. The combined increased frequency of storms and sea level rise in response to global warming will contribute to coastal erosion, especially where there are cliffs composed of soft rocks such as in Pacifica, located along the Pacific Ocean coast in California, and Hapsburg, located along the North Sea in East Anglia, England. Erosion problems can be worsened by countermeasures that do not consider the effects on adjacent shores. Giardino et al. [2018] discussed erosion along low-lying islands along the south eastern side of Kwajalein Atoll.

Estuary: An embayment of the coast such as a river, bay, or lagoon where fresh river water entering at its head mixes with the relatively saline ocean water. A North American example is the St. Lawrence River, which connects the Great Lakes to the Atlantic Ocean. Increasing water levels and salinity changes can be very damaging to an estuary. Plants that are meant to be above the waterline such as least spikerush (*Eleocharis acicularis*) are now being drowned, and there is ever-decreasing light availability to submerged aquatic vegetation such as eelgrass (*Zostera marina*). Sea level rise also increases the amount of flooding and erosion of coastal areas. Inundation and saltwater intrusion increase the salinity of water that was once fresh.

Eustatic change: Global fluctuations in sea level due to an alteration in the volume of water in the oceans or, alternatively, a change in the amount of water owing to a change in the shape of an ocean basin. A eustatic sea level rise refers to an increase in the volume of the world's waters caused by the melting of polar ice caps since the last ice age ended 11,700 years ago.

Evaporation: The process by which water changes from a liquid to a gas or vapor. Hypersaline conditions in estuaries such as the Laguna Madre in the United States (e.g., Kolker [2003]) or the Coorong Estuary, Leschenault Estuary, and Spencer Gulf in Australia (e.g., Wolanski [2014]) occur as salinity levels rise during the dry summer when higher temperatures increase levels of evaporation in the estuary.

Extra-tropical cyclone: A cyclonic-scale storm that primarily gets its energy from the release of potential energy when cold and warm air masses interact. These storms always have one or more fronts connected to them, and can occur over the land or ocean. The tracks of these storms affect weather in mid-latitudes. Extratropical cyclones include blizzards, Nor'easters, and the ordinary low-pressure systems that give the continents at mid-latitudes much of their precipitation. Some research meteorologists have provided evidence of an increase in the frequency of winter storms and their intensity since the 1950s, and their tracks have shifted northward over the United States. Extra-tropical cyclones can produce a storm surge and cause coastal erosion.

Glacier: A large, perennial accumulation of ice, snow, rock, sediment, and liquid water originating on land and moving down slope under the influence of its own weight and gravity. Glaciers are classified by their size, location, and thermal regime and may terminate on land or in water. One of the largest glaciers in the world is the Lambert–Fisher Glacier in Antarctica. Melting glaciers from global warming cause increased runoff which contributes to sea level rise.

Greenhouse gases: Any gas that absorbs infrared radiation in the atmosphere. Greenhouse gases include, carbon dioxide, chlorofluorocarbons, hydrochlorofluorocarbons, hydrofluorocarbons, methane, nitrous oxide, ozone, perfluorocarbons, sulfur hexafluoride, and soot. The construction industry is considered a substantial contributor to greenhouse gas emissions. Giesekam and Barrett [2014] discussed possible strategies to reduce greenhouse gases through improved building design, increased recycling, supply chain efficiency measures, and adaptive reuse.

Harmful algal blooms or HABs: A rapid increase in the population of algae in an aquatic or marine environment. Very dense clouds of these organisms (blooms) can change the color of the water to green, yellowish-brown, or red. Bright green blooms may also occur as a result of blue-green algae, which are actually bacteria (cyanobacteria). Only a few dozen of the many thousands of species of microscopic and macroscopic algae (phytoplankton) are repeatedly associated with toxic or harmful blooms. Species, such as dinoflagellates (e.g., *Alexandrium tamarense*, *Akashiwo sanguinea*, *Karenia brevis*, and *Pfiesteria piscicida*) and the diatom *Pseudo-nitzschia australis*, produce potent toxins which are liberated when the algae are eaten. Harmful algal blooms can harm or kill marine animals, contaminate shellfish, and threaten human life.

High marsh: A sandy region of transition between low marsh and maritime forest zones. The high marsh is only flooded during the highest of spring tides and storm surges. Its higher elevation and diverse suite of plant species including grasses, wildflowers, and shrubs provide a natural buffer to intercept and filter stormwater runoff and groundwater flowing from upland areas.

Hurricane: An intense tropical cyclone in which winds tend to spiral inward toward a core of low pressure, with maximum wind velocities that are greater than or equal to 33.5 m/s (75 mph) for several minutes or longer. The term hurricane is used in the Atlantic Ocean, Gulf of Mexico, and eastern Pacific Ocean. As an example, tropical depression Harvey moved into the warm waters of the Gulf of Mexico on August 22, 2017, and exploded rapidly into a major hurricane in approximately 40 hr. Harvey made landfall as a Category 4 Hurricane with winds of 209 km/hr (130 mph) near Rockport, Texas on August 25, 2017. Hurricane Harvey dropped 1–1.5 m (40–61 in.) of rainfall in southeast Texas and southwest Louisiana and created a catastrophic flood disaster in southeast Texas.

Hydrologic cycle: The process of evaporation, vertical and horizontal transport of vapor, condensation, precipitation, and the flow of water from continents to oceans. It is a major factor in determining climate through its influence on surface vegetation, the clouds, snow and ice, and soil moisture. The hydrologic cycle is responsible for 25–30% of the mid-latitudes' heat transport from the equatorial to polar regions.

Hydrosphere: All the waters on the earth's surface including clouds, oceans, seas, rivers, lakes, underground water, etc. Like the atmosphere, the hydrosphere is in constant movement through land drainage from rivers and streams and circulation in estuaries and ocean currents. Ocean currents are particularly important in transferring heat from the equator toward the poles.

Hypoxia: Low or depleted oxygen in coastal ocean and freshwater environments that is associated with the overgrowth of algae, which can lead to oxygen depletion when they die, sink to the bottom, and decompose. These areas of low dissolved oxygen concentration are called "dead zones" because animals can suffocate and die. Dead zones caused by agricultural run-off and other pollutants are found all around the world. Examples include Lake Erie, Chesapeake Bay, Lake Tai in China, the Adriatic Sea, Baltic Sea, Black Sea, and Gulf of Mexico. Researchers studying hypoxia in the Great Lakes have described physical processes causing hypoxia in Lake Erie [Rao et al., 2008] and experimental forecasts are made available by the Great Lakes Environmental Research Laboratory.

Impervious surface: Surfaces such as roofs, solid decks, driveways, patios, sidewalks, parking areas, tennis courts, and concrete or asphalt streets that impede the natural infiltration of water into the soil. Since impervious surfaces seal the soil surface, eliminating rainwater infiltration and natural groundwater recharge, they contribute to increased surface runoff and flooding. Elvidge et al. [2004] discussed the impact of built-up areas in the United States on local hydrology, climate, and carbon cycling.

Inferential statistics: Techniques that support conclusions, generalizations, and predictions from observations. One might evaluate observations to determine whether there is correlation or causation among factors such as wave steepness and beach erosion that impacts the landfall points for submarine telecommunications cables. Sampson et al. [2016] described a Monte Carlo wind speed probability algorithm that was used in tropical cyclone forecasting.

Inundation: The submergence of land by water, particularly in a coastal setting. The effect, result, or outcome of inundation is potential loss of life, economic losses, and adverse social-environmental impacts. Inundation caused by storms may be measured by water level gages, which provide important information on how high-water levels rise and their duration at a particular location. This type of data has proven useful for communities that are interested in preparing for sea level rise. Cities such as Annapolis, Maryland in the United States, located at the confluence of the Severn River and Chesapeake Bay, have flooded in the past and continue to be at risk of flooding. Historic coastal structures in Annapolis such as the Sands House (c. 1740) are considered to be highly vulnerable to future flooding.

Invasive species: Any kind of living organism (e.g., bacteria, fungus, plant, insect, mollusk, fish, reptile, or mammal) that is not native to an ecosystem and which causes harm. Most commercial, agricultural, and recreational activities depend on healthy native ecosystems. For this reason, the impacts of invasive species on natural ecosystems and the economy costs billions of dollars each year. In the United States, costly effects include clogging of water facilities from quagga (*Dreissena bugensis*) and zebra (*Dreissena polymorpha*) mussels and clogging of waterways from aquatic plants such as the weed hydrilla (*Hydrilla verticillata*) and giant salvinia (*Salvinia molesta*), disease transmission, harm to fisheries, and increased fire vulnerability and diminished grazing value.

Isostasy: The balance between changes within the Earth's crust and mantle, where material is displaced in response to an increase (isostatic depression) or decrease (isostatic rebound) in mass at any point on the Earth's surface above. Such changes are frequently caused by advances or retreats of glaciers. Govers [2009] theorized that isostasy sealed off the Mediterranean Sea from the Atlantic Ocean five million years ago.

Isostatic change: Local fluctuations in sea level in response to an increase or decrease in the height of the land. When the height of the land increases in a place such as Richmond Gulf in southeastern Hudson Bay, the sea level falls and when the height of the land decreases in a place such as the Chesapeake Bay the sea level rises. In addition to the sinking of land that was elevated at the edge of a glacier, subsidence may be caused by non-geological processes. For example, land subsidence in the Houston–Galveston, Texas, area and in the Santa Clara Valley, California area were caused by petroleum extraction in Texas and groundwater withdrawals, as well as ground water withdrawals in California. For these reasons, isostatic change refers to local sea level changes while eustatic change refers to global sea level change.

Lagoon: A shallow water body found on all continents which receives little if any fluvial input and is connected to the ocean by coastal inlets. Extreme storms causing the inundation of lagoons by the coastal ocean will change the salinity and possibly alter the ecosystem. Salinity can have a great impact on the type of organisms that live in a body of water such as the Pamlico Sound in North Carolina, the Laguna Madre in Texas, the Laguna Madre de Tamaulipas in Mexico, Lake Nokoué in Benin, and Lake Piso in Liberia.

Landfast ice: A type of sea ice that is anchored to shore, grounded icebergs, or the shallow seafloor. It does not move with the winds or currents. Landfast ice is particularly important to polar bears (*Ursus maritimus*) and provides access to prey such as ringed seal (*Pusa hispida*) pups. Dammann et al. [2018] described the use of synthetic aperture radar interferometry to provide quantitative information for ice road planning, construction, and maintenance.

Land and sea breezes: Breezes occurring near shorelines that are caused by unequal heating of air over land and water. While the land is warm during the day, air above it rises, and a cool breeze blows in from the sea. As the land cools off at night, air pressure over it increases, and a cool land breeze blows out to the sea. The leading edge of a sea breeze or sea breeze front can provide a trigger to daily coastal thunderstorms.

Land subsidence: The gradual settling or sudden sinking of the Earth's surface owing to subsurface movement of earth materials. Coastal and delta cities around the world are especially susceptible to the sinking of land from the extraction of groundwater, oil, and gas and the drainage of soils [Showstack, 2014].

La Niña: Coupled ocean-atmospheric interactions across a broad expanse of the tropical Pacific Ocean with a global impact on weather patterns. The condition refers to the extensive cooling of the central and eastern tropical Pacific Ocean, often accompanied by warmer than normal sea surface temperatures in the western Pacific, and to the north of Australia. La Niña events are sometimes thought of as the opposite of El Niño.

Levee: An artificial bank confining a river or stream channel or limiting adjacent areas subject to flooding. Levees help protect people in the Netherlands from flooding by the North Sea, in Canada from flooding by the Bay of Fundy, and in the United States from flooding by the Mississippi river. Earthen levees must be maintained owing to erosion, especially since failure at one location can result in a catastrophic failure for an entire levee system. Jadid et al. [2020] have suggested that repeated flooding events from events such as Hurricanes Floyd and Matthew have a cumulative effect on the structural integrity of earthen levees.

Living shorelines: Shoreline protection that provides erosion control benefits through the strategic placement of plants, stone, sand fill, and organic structural materials such as "biologs" and oyster reefs. Biologs are made of biodegradable (e.g., coconut fiber) materials bound by high strength twisted netting and may be staked to an eroding shoreline to help attenuate wave energy. Oyster larvae typically settle on the shells of other oysters, forming dense, expansive clusters, or colonial communities known as oyster reefs, bars, or beds. They may also settle on blocks of concrete, limestone, crushed shell, and silica. Mitchell and Bilkovic [2019] discussed siting factors, design considerations, and maintenance for living shorelines.

Longshore current: Nearshore currents occurring as a result of incident waves that approach the shoreline at an angle. The flow is parallel to the shore and extends from the shoreline out through the surf zone. Harrison [1968] provided an empirical equation for longshore current velocity based on factors such as beach slope, and the breaker height, period, angle, crest length, and trough depth.

Low marsh: A silt laden region that is flooded daily and characterized by vegetation such as Spartina alterniflora and some types of algae. Restoration efforts may involve the installation of low-profile structures such as marsh sills to contain fill for newly planted marshes. Sills should be designed to match the expected wave climate.

Macro algae: Large marine and aquatic photosynthetic plants (seaweed) that can be seen without the aid of a microscope. Macro algae are ecologically and economically important primary producers and play key roles in coastal carbon cycles. These sessile organisms are impacted by factors such as land runoff, water pollution, and increasing water temperatures. Macroalgae beds (e.g., *Ascophyllum nodosum*, *Gracilaria pacifica*, *Fucus gardneri*, and *Sargassum muticum*) provide important ecosystem goods and services and the degradation of these systems will have far reaching consequences to society.

Monsoon: A thermally driven wind arising from differential heating between a land mass and the adjacent ocean that reverses its direction seasonally. The India monsoon is caused by the asymmetric heating and cooling of the Indian Ocean during summer and winter. It blows from the northeast during cooler months and reverses direction to blow from the southwest during the warmest months of the year. This process brings large amounts of rainfall to India during June and July.

Morphodynamics: The study of landscape and seascape changes due to erosion and sedimentation. Rising sea level, increases in storm intensity, and storm frequency increase the magnitude of forces which impact complex and impermanent coastal systems. Combinations of high-resolution, high-precision spatiotemporal, process-based field observations with numerical models are being applied to predict the evolution of the estuaries, beaches, tidal inlets, and hydrography of the coastal zone.

Mud: Wet clay and silt-rich sediment. Navigation channels may become too shallow for navigation owing to sedimented sand and mud, requiring operations such as maintenance dredging. The ongoing maintenance of navigation channels helps the world economy by promoting efficient trade via waterways.

Multispectral imagery: A multi-channel raster consisting of several spectral bands of wavelengths such as red, green, blue and near infrared. Multispectral images may be used to assess shoreline change, which provides important information to protect coastal properties and preserve coastal environments. Brodie et al. [2018] described procedures to map submerged aquatic vegetation along the Thanet Coast in south-east England.

Neap tides: Tides of decreased range or tidal currents of decreased speed occurring semimonthly as the result of the moon being in quadrature. During neap tides, water levels may increase when combined with flood drivers such as storm surge or river discharge.

Northeasters: An extratropical cyclone with winds blowing from the northeast. In parts of the United States such as New England, these storms are called, "nor'easters." A 22 m/s (50 mph) nor'easter during January 1987 combined with a high tide cut a new inlet through North Beach in Cape Cod, Massachusetts forming what the locals call, "South Beach Island." The new inlet has narrowed in recent years as the southern tip of North Beach Island has lengthened, pushing the coastal inlet against an eroding South Beach.

Nutrients: Substances that provide nourishment for growth or metabolism. Plants absorb *nutrients* and animals obtain *nutrients* from ingested foods. Too many nutrients entering coastal waters (e.g., nitrogen and phosphorus) will cause an excessive growth or bloom of algae followed by low levels of dissolved oxygen.

Ocean acidification: Increased concentrations of carbon dioxide in sea water causing a measurable increase in acidity (i.e., a reduction in ocean pH). This will impact calcifying organisms such as corals and mollusks.

Perigean spring tide: A tide occurring when the moon is either new or full and closest to Earth (perigee). Perigean spring tides may cause minor coastal flooding in very low-lying areas such as Fort Lauderdale, Florida. It is expected that occurrences of minor "nuisance" flooding at the times of perigean spring tides will increase as sea levels rise relative to the land.

Recession: Landward movement of the shoreline over a specified period of time. Rates of shoreline recession are higher on beaches exposed to high wave energy.

Relative sea level rise: The increase in ocean water levels at a specific location, taking into account both global sea level rise and local factors, such as local subsidence and uplift. Relative sea level rise is measured with respect to a specified vertical datum relative to the land, which may also be changing elevation over time.

Resilience: A capability to anticipate, prepare for, respond to, and recover from significant multi-hazard threats with minimum damage to social well-being, the economy, and the environment.

Rip current: A strong, narrow, and nearshore current which flows directly from shore through the surf zone. Castelle et al. [2016] provided a review on rip current processes and associated hazards including drowning deaths.

Rogue wave: Large, unexpected, and dangerous waves that are twice the significant wave height of the surrounding waves. Dudley et al. [2019] review rogue wave physics in the ocean and in optical fibers. Didenkulova [2020] has analyzed and described 210 hazardous rogue wave events and regions based on data collected from 2011–2018.

Saffir–Simpson scale: A 1–5 rating based on a hurricane's sustained wind speed and potential property damage. Hurricanes reaching Category 3 and higher are considered major hurricanes because of their potential for significant loss of life and damage. Category 1 and 2 storms are still dangerous, however, and require preventative measures. The scale was originally developed by civil engineer Herb Saffir 1917–2007 and meteorologist Bob Simpson 1912–2014. It has been an excellent tool for alerting the public about the possible impacts of various intensity hurricanes.

Salinity: The dissolved salt content of a body of water. Marine salinity levels are influenced by a number of factors including rainfall, evaporation, inflow of river water, wind, and melting of glaciers. Salinity plays a critical role in the water cycle and ocean circulation.

Sand: Loose particles of rock or mineral (sediment) that range in size from 0.0625–2.0 mm in diameter. Sands are common type of marine sediment that are transported alongshore to produce sandy beaches.

Saltwater intrusion: Displacement of fresh or ground water by the advance of saltwater due to its greater density, usually in coastal and estuarine areas.

Sea level rise: The upward trend in average sea level height. Causes of sea level rise include the thermal expansion of seawater by warming of the ocean and increased melting of land-based ice, such as glaciers and ice sheets. Higher sea levels allow floods to push farther inland and future rising sea level impacts may be foreshadowed by recurring flooding occurring during spring tides. NOAA scientists Sweet et al. [2014] explained sea level rise, nuisance flooding, and risks to coastal roads, seaports, and even cities based on water level gage data.

Sinkhole: A depression in the surface commonly found in in karst landscapes. Sinkholes often form where limestone or some other soluble rock is partially dissolved by groundwater, then collapses to form a depression.

Spring tide: Tides of increased range or tidal currents of increased speed occurring semimonthly as the result of the moon being new or full.

Storm surge: A storm-induced rise of water, over and above the predicted astronomical tides. The wind-induced surge can cause extreme flooding to low-lying coastal areas. Menéndez et al. [2020] describe coastal flooding and the economic value of natural barriers such as mangrove forests that reduce flood risk.

Subsidence: Land sinking as the result of multiple causes including compaction of coastal sediments, sinking under the shear weight of large urban centers, withdrawal of groundwater or oil, and regional tectonic effects. Subsidence is added to, and exacerbates, sea level rise. Most of the world's river deltas are experiencing subsidence.

Thunderstorm: A rain-bearing cloud that also produces lightning. Every thunderstorm produces lightning, which may have negative impacts on engineering projects. Cartier [2017] described advances in using historic storm data and ongoing storm conditions to predict lightning.

Tidal current: The horizontal movement of water resulting from gravitational effects of the Earth, Sun, and Moon, without any atmospheric influences.

Tide: The vertical fluctuations in water level resulting from gravitational effects of the Earth, Sun, and Moon, without any atmospheric influences. As sea levels rise high tides will cause seawater to reach further inland. In some locations, tidal flooding is causing erosion, aquifer and agricultural soil contamination, and lost habitat for fish, birds, and plants.

Tornado: A violently rotating column of air, usually pendant to a cumulonimbus, with circulation reaching the ground. It nearly always starts as a funnel cloud and may be accompanied by a loud roaring noise. On a local scale, it is the most destructive of all atmospheric phenomena.

Tropical cyclone: A warm-core non-frontal synoptic-scale cyclone, originating over tropical or subtropical waters, with organized deep convection and a closed surface wind circulation about a well-defined center. Once formed, a tropical cyclone is maintained by the extraction of heat energy from the ocean at high temperature and heat export at the low temperatures of the upper troposphere. In this they differ from extratropical cyclones, which derive their energy from horizontal temperature contrasts in the atmosphere (baroclinic effects).

Troposphere: The lowest part of the atmosphere from the surface to about 10 km (6.2 mi) in altitude in mid-latitudes (ranging from 9 km (5.6 mi) in high latitudes to 16 km (9.9 mi) in the tropics on average) where clouds and "weather" phenomena occur.

Tsunami: A long-period wave caused by an underwater disturbance such as a volcanic eruption or earthquake. Schnyder et al. [2016] have provided evidence for large submarine landslides on the slopes of the Great Bahama Bank in the Caribbean that have generated tsunamis in the past.

Typhoon: A tropical cyclone in the western Pacific Ocean. Typhoon Tip (aka Typhoon Warling) is an example of a typhoon that rapidly intensified over the open waters of the western Pacific Ocean around October 10, 1979, and then began to weaken steadily after passing east of Okinawa and finally making landfall on the Japanese island of Honshū with winds of about 81 mph (130 km/hr) on October 19, 1973. Heavy rainfall from the typhoon caused severe flooding and contributed to a devastating fire at Camp Fuji, a U.S. Marine Corps training facility located in the Shizuoka Prefecture of Japan, at the base of Mount Fuji.

Upwelling: The process where favorable winds blowing across the ocean surface push water away which allows deep, cold, and nutrient rich water to rise up from beneath the surface to replace the water that was pushed away. Upwelling generally occurs in the ocean along coastlines. Two important upwelling centers that vary in strength and frequency occur along the coast from Washington to Southern California in North America and along the coast from Peru to Chile in South America.

Water level: Elevation of still water level to some datum. Water levels fluctuate as a result of astronomical tides, basin oscillations, climatological effects, evaporation and precipitation, geological effects, storm surge, and tsunami. The water level dictates where waves can reach and attack the coastal zone.

Wave: A ridge, deformation, or undulation of the surface of a liquid. Understanding the interaction of waves with maritime operations such as docking ships, building and maintaining coastal structures, and protecting navigation channels and beaches is critical for coastal resilience.

Weather: The day-to-day state of the atmosphere as regards to sunshine, temperature, humidity, precipitation, wind, cloudiness, moisture, pressure, etc. Weather forecasters apply technology (e.g., meteorological satellites and Doppler radar) and science to predict the state of the atmosphere for a future time and at a given location. This includes collecting as much data as possible and using these observations as input to numerical computer models through a process known as data assimilation to produce outputs of meteorological elements from the surface of the ocean to the top of the atmosphere.

Wind waves: Waves being formed and built by the wind. Wave height is affected by wind speed, wind duration (how long the wind blows), and fetch (the distance over which the wind blows).

Wind stress: The way the faster moving wind transfers energy to the slower moving water. Wind stress drives currents and forces waves and storm surges, which leads to flooding during the passage of storms such as hurricanes. Many researchers have found that surface currents are around 3% the wind velocity at 10 m above the sea surface [Open University, 2001, Weber, 1983].

Bibliography

Amaral, S. S., de Carvalho, J. A., Costa, M. A. M., and Pinheiro, C. (2015). An overview of particulate matter measurement instruments, *Atmosphere*, 6(9):1327–1345. DOI: 10.3390/atmos6091327. 20

American Meteorological Society, (2020). *Glossary of Meteorology*. http://glossary.ametsoc.org/wiki/Main_Page

Arabi, B., Salama, M. S., Pitarch, J., and Verhoef, W. (2020). Integration of in-situ and multi-sensor satellite observations for long-term water quality monitoring in coastal areas, *Remote Sensing of the Environment*, 239(111632). DOI: 10.1016/j.rse.2020.111632. 42

Aroucha, L. C., Duarte, H. O., Droguett, E. L., and Veleda, D. R. A. (2018). Practical aspects of meteorology and oceanography for mariners: A guide for the perplexed, *Cogent Engineering*, 5(1), Article 1492314. DOI: 10.1080/23311916.2018.1492314. 12

Bachmann, C. M., Ainsworth, T. L., Fusina, R., Montes, M., Bowles, J., Korwan, D., and Gillis, D. (2009a). Bathymetric retrieval from hyperspectral imagery using manifold coordinate representations, *IEEE Transactions on Geoscience and Remote Sensing*, 47(3):884–897. DOI: 10.1109/tgrs.2008.2005732. 9

Bachmann, C. M., Nichols, C. R., Montes, M. J., Li, R.-R., Woodward, P. K., Fusina, R. A., Chen, W., Mishra, V., Kim, W., Monty, J., McIlhany, K., Kessler, K., Korwan, D., Miller, D., Bennert, E., Smith, G., Gillis, D., Sellars, J., Parrish, C., Weidemann, A., Goode, W., Schwarzschild, A., and Truitt, B. (2009b). Airborne remote sensing of trafficability in the coastal zone. *NRL Review*, pp. 223–228, Naval Research Laboratory, Washington, DC. 53

Bachmann, C. M., Montes, M. J., Fusina, R. A., Parrish, C., Sellars, J., Weidemann, A., Goode, W., Nichols, C. R., Woodward, P., McIlhany, K., Hill, V., Zimmerman, R., Korwan, D., Truitt, B., and Schwarzschild, A. (2010a). Bathymetry retrieval from hyperspectral imagery in the very shallow water limit: A case study from the 2007 Virginia Coast Reserve (VCR'07) multi-sensor campaign, *Marine Geodesy*, 33(1):53–75. DOI: 10.1080/01490410903534333. 9, 47

Bachmann, C. M., Nichols, C. R., Montes, M, Li, R., Woodward, P., Fusina, R. A., Chen, W., Mishra, V., Kim, W., Monty, J., McIlhany, K., Kessler, K., Korwan, D., Miller, D., Bennert, E., Smith, G., Gillis, D., Sellers, J., Parrish, C., Schwarzschild, A., and Truitt, B.

(2010b). Retrieval of substrate bearing strength from hyperspectral imagery during the Virginia Coast Reserve (VCR '07) multi-sensor campaign, *Marine Geodesy*, 33(2–3):101–116. DOI: 10.1080/01490419.2010.492278. 47

Bachmann, C. M., Fusina, R. A., Montes, M. J., Li, R-R., Gross, C., Nichols, C. R., Fry, J. C., Parrish, C., Sellars, J., White, S., Jones, C., and Lee, K. (2012a). Talisman-Saber 2009 remote sensing experiment, *NRL Memorandum Report NRL/MR/7230–9404*, Naval Research Laboratory, Washington, DC. 18, 47, 53

Bachmann, C. M., Fusina, R. A., Montes, M. J., Li, R-R., Nichols, C. R., Fry, J. C., Hallenborg, E., Parrish, C. E., and Sellars J. (2012b). Hawaii-hyperspectral airborne remote environmental sensing (HIHARES'90) experiment, *NRL Memorandum Report NRL/MR/7230–12-9403*, Naval Research Laboratory, Washington, DC. DOI: 10.21236/ada559833. 47, 53

Bachmann, C. M., Fusina, R. A., Montes, M. J., Li, R-R., Gross, C., Fry, J. C., Nichols, C. R., Ezrine, D. F., Miller, W. D., Jones, C., Lee, K., Wende, J., and McConnon, C. (2012c). Mariana Islands–hyperspectral airborne remote environmental sensing experiment 2010, *NRL Memorandum Report NRL/MR/7230–12–9405*, Naval Research Laboratory, Washington, DC. DOI: 10.21236/ada559525. 47, 53

Bachmann, C. M., Abelev, A., Bounds, M. T., Nichols, C. R., Fusina, R. A., Li, R-R., Kremens, R., Difrancesco, G., Vermillion, M., Mattis, G., Edwards, K. B., and Estrada, J. (2016). Ssang Yong 2014 remote sensing experiment, *NRL Memorandum Report, NRL/MR/7230–16–9667*, Naval Research Laboratory, Washington, DC. 53

Bachmann, C. M., Eon, R. S., Lapszynski, C. S., Badura, G. P., Vodacek, A., Hoffman, M. J., McKeown, D., Kremens, R. L., Richardson, M., Bauch, T., and Foote, M. (2018). A low-rate video approach to hyperspectral imaging of dynamic scenes, *Journal of Imaging*, 5(1), 6. DOI: 10.3390/jimaging5010006. 47

Bachmann, C. M. and Nichols, C. R. (2019). Hyperspectral imager for updated littoral situational awareness (HULA) [day 1 presentation]. *NOAA Emerging Technologies Workshop, Extreme Weather and Water, Report*, College Park, MD. https://nosc.noaa.gov/public_docs/NOAA_2019_ETW_report.pdf 47

Baker, M. (2016). Quality time: It may not be sexy, but quality assurance is becoming a crucial part of lab life, *Nature*, 529:456–458. 39

Ball, M. (2019, April 28). TigerShark-XP UAVs receive FAA experimental certification, unmanned systems news. https://www.unmannedsystemstechnology.com/2019/04/tigershark-xp-uavs-receive-faa-experimental-certification/ 7

Ballard, R. D., Raineault, N. A., Fahey, J., Mayer, L., Heffron, E., Broad, K., Bursek., J., Roman, C., and Krasnosky, K. (2018). Submerged sea caves of Southern California's continental boarderland, *Oceanography,* 31(1), Supplement, 30–31. 9

Beiras, R. (2018). *Marine Pollution: Sources, Fate and Effects of Pollutants in Coastal Ecosystems.* Amsterdam, The Netherland, Elsevier. 19

Berry, D. I. and Kent, E. C. (2009). A new air-sea interaction gridded dataset from ICOADS with uncertainty estimates, *Bulletin of the American Meteorological Society,* 90(5):645–656. DOI: 10.1175/2008bams2639.1. 3

Berteaux, H. O. (1976). *Buoy Engineering,* New York, John Wiley & Sons. 38

Bidlot, J-R. and Holt, M. W. (2006). Verification of operational global and regional wave forecasting systems against measurements from moored buoys, *JCOMM Technical Report: 30,* WMO and IOC, 11pp. (WMO TD: 1333), Geneva, Switzerland. http://hdl.handle.net/11329/101 31

Bigorre, S., Suhm, D., Donohue, M., Franks, A., Kuo, J., Wellwood, D., Schwartz, J., and Kelly, B. (2017). Argentine basin 3 deployment cruise report, RVIB Palmer NBP 16–09, October 20 November 15, 2016, Punta Arenas, Chile. Control no. 3206-00303. Woods Hole, MA: Woods Hole Oceanographic Institution. 6

Bird, E. and Lewis, N. (2015). *Beach Renourishment.* Dordrecht, The Netherlands, Springer. DOI: 10.1007/978-3-319-09728-2. 57

Bishop, J. M. (1984). *Applied Oceanography.* New York, John Wiley & Sons. 1

Bloetscher, F., Sham, C. H., Danko, J. J., and Ratick, S. (2014). Lessons learned from aquifer storage and recovery (ASR) systems in the United States, *Journal of Water Resources and Protection,* 6(17):1603–1629. DOI: 10.4236/jwarp.2014.617146. 56

Boyd, M. B. and Sommerfield, C. K. (2016). Marsh accretion and sediment accumulation in a mangaged tidal wetland complex of Delaware Bay, *Ecological Engineering,* 92:37–46. DOI: 10.1016/j.ecoleng.2016.03.045. 55

Bretschneider, C. L. (1959). Wave variability and wave spectra for wind generated gravity waves, *Technical Memo No. 118,* Washington, DC, Beach Erosion Board, U.S. Army Corps of Engineering. 44

Brodie, J., Ash, L. V., Tittley, I., and Yesson, C. (2018). A comparison of multispectral aerial and satellite imagery for mapping intertidal seaweed communities, *Aquatic Conservation Marine and Freshwater Ecosystems,* 28(4):872–881. DOI: 10.1002/aqc.2905. 67

Bueger, C. (2015). What is maritime security?, *Marine Policy*, 53:159–164. DOI: 10.1016/j.marpol.2014.12.005. 12

Caires, S. (2011). Extreme value analysis: Wave data, *JCOMM Technical Report 57*, Geneva, Switzerland, World Meteorological Organization. https://www.jodc.go.jp/info/ioc_doc/JCOMM_Tech/JCOMM-TR-057.pdf 29

Caldwell, P. C. and Aucan, J. P. (2007). An empirical method for estimating surf heights from deepwater significant wave heights and peak periods in coastal zones with narrow shelves, steep bottom slopes, and high refraction, *Journal of Coastal Research*, 23(5):1237–1244. DOI: 10.2112/04-0397r.1. 25

Capelotti, P. J. (1996). *Oceanography in the Coast Guard*. Washington, DC, U.S. Coast Guard Historians Office. 12

Cartier, K. M. S. (2017). New model predicts lightning strikes; alert system to follow, *Eos*, 98. DOI: 10.1029/2017eo088591. 69

Castelle, B., Scott, T., Brander, R. W., and McCarroll, R. J. (2016). Rip current types, circulation and hazard, *Earth-Science Reviews*, 163:1–21. DOI: 10.1016/j.earscirev.2016.09.008. 68

Center for Operational Oceanographic Products and Services, (2000). *Tide and Current Glossary*, Silver Spring, MD, NOAA.

Coleman, J. and Huh, O. K. (2004). Major world deltas: A perspective from space. *World Delta Database*. www.geol.lsu.edu/WDD 60

Corbett, J. J. and Fischbeck, P. S. (1997). Emissions from ships, *Science*, 278:823–824 DOI: 10.1126/science.278.5339.823. 20

Corbett, J. J. and Fischbeck, P. S. (2000). Emissions from waterborne commerce vessels in United States continental and inland waterways, *Environmental Science and Technology*, 34:3254–3260. DOI: 10.1021/es9911768. 20

Cotton, C. A. (1954). Deductive morphology and the genetic classification of coasts, *Science Monthly*, 78(3):163–181. 13

Cox, A. T. and Swail, V. R. (2001). A global wave hindcast over the period 1958—vol. 1997: Validation and climate assessment, *Journal of Geophysical Research: Oceans*, 106(C2):2313-2329. DOI: 10.1029/2001jc000301. 10

Cummins, P. F. and Pramod, T. (2018). A note on evaluaring model tidal currents against observations, *Continental Shelf Research*, 152:35–37. DOI: 10.1016/j.csr.2017.10.007. 23

Dammann, D. O., Eicken, H., Mahoney, A. R., Meyer, F. J., Freymueller, J. T., and Kaufman, A. M. (2018). Evaluating landfast sea ice stress and fracture in support of operations on sea ice using SAR interferometry, *Cold Regions Science and Technology*, 149:51–64. DOI: 10.1016/j.coldregions.2018.02.001. 65

De Mey, P., Craig, P., Davidson, F., Edwards, C. A., Ishikawa, Y., Kindle, J. C., Proctor, R., Thompson, R. K., and Zhu, J. (2009). Applications in coastal modeling and forecasting, *Oceanography*, 22(3):198–205. DOI: 10.5670/oceanog.2009.79. 10

Debroas, D., Mone, A., and Ter Halle, A. (2017). Plastics in the North Atlantic garbage patch: A boat-microbe for hitchhikers and plastic degraders, *Science of the Total Environment*, 599–600, 1222–1232. DOI: 10.1016/j.scitotenv.2017.05.059. 21

Defant, A. (1961). *Physical Oceanography*, 2, New York, Pergamon Press. 22

Dickey, T. D. (1991). The emergence of concurrent high-resolution physical and bio-optical measurements in the upper ocean and their applications, *Reviews of Geophysics*, 29(3):383–413. DOI: 10.1029/91rg00578. 42

Dickey, T. D. and Bidigare, R. R. (2005). Interdisciplinary oceanographic observations: The wave of the future, *Scientia Marina*, 69:23–42. DOI: 10.3989/scimar.2005.69s123. 1, 2

Dickson, A. G. (2010). Standards for ocean measurements, *Oceanography*, 23(3):34–47. DOI: 10.5670/oceanog.2010.22. 39

Didenkulova, E. (2020). Catalogue of rogue waves occurred in the World Ocean from 2011 to 2018 reported by mass media sources, *Ocean and Coastal Management*, 188(105076). DOI: 10.1016/j.ocecoaman.2019.105076. 68

Doodson, A. T. and Warburg, H. D. (1941). *Admiralty Manual of the Tides*. London, His Majesty's Stationery Office. 21

Dorwart, J. M. (1992). *Cape May County, New Jersey: The Making of an American Resort Community*. New Brunswick, NJ, Rutgers University Press. 34

Dudley, J. M., Genty, G., Mussot, A., Chabchoub, A., and Dias, F. (2019). Rogue waves and analogies in optics and oceanography, *Nature Reviews Physics*, 1:675–689. DOI: 10.1038/s42254-019-0100-0. 68

Dunkin, L. M., Coe, L. A., and Ratcliff, J. J. (2018). Corps shoaling analysis tool: Prediction channel shoaling, ERDC/CHL TR-18–16, Vicksburg, MS, Engineering Research and Development Center, U.S. Army Corps of Engineers. DOI: 10.21079/11681/30382. 50

Dybas, C. L. (2005). Dead zones spreading in world oceans, *BioScience*, 55(7):552–557. DOI: 10.1641/0006-3568(2005)055[0552:dzsiwo]2.0.co;2. 20

Edwards, K. (2011). Ocean testing of a power-capturing wave buoy [Session 1]. *Ocean Waves Workshop*, University of New Orleans, New Orleans, LA. https://scholarworks.uno.edu/cgi/viewcontent.cgi?article=1000&context=oceanwaves 45

Elvidge, C. D., Milesi, C., Dietz, J. B., Tuttle, B. T., Sutton, P. C., Nemani, R., and Volgemann, J. E. (2004). U.S. constructed area approached the size of Ohio, *Eos, Transactions, American Geophysical Union*, 85(24):233–240. DOI: 10.1029/2004eo240001. 64

Emery, W. J. and Thomson, R. E. (2014). *Data Analysis Methods in Physical Oceanography*, 3rd ed., Amsterdam, Netherlands, Elsever Science. DOI: 10.1016/B978-0-444-50756-3.X5000-X. 1

Eon, R. S., Bachmann, C. M., Lapszynski, C. S., Tyler, A. C., and Goldsmith, S. (2020). Retrieval of sediment filling factor in a salt panne from multi-view hyperspectral imagery, *Remote Sensing*, 12(3):422. DOI: 10.3390/rs12030422. 47

Fairbridge, R. W. (2004). Classification of coasts, *Journal of Coastal Research*, 20(1):155–165. DOI: 10.2112/1551-5036(2004)20[155:coc]2.0.co;2. 13

Ferrersen, F. and Trowbridge, J. H. (2005). The effect of wave breaking on surf-zone turbulence and alongshore currents: A modeling study, *Journal of Physical Oceanography*, 35(11):2187–2203. DOI: 10.1175/jpo2800.1. 58

Finkl, C. W. (2004). Coastal classification: Systematic approaches to consider in the development of a comprehensive scheme, *Journal of Coastal Research*, 20(1):166–213. DOI: 10.2112/1551-5036(2004)20[166:ccsatc]2.0.co;2. 13

Foreman, M. G. G. (1978). Manual for tidal currents analysis and prediction. Pacific Marine Science Report 78–6, Patricia Bay, Sidney, BC, Institute of Ocean Sciences. http://www.dfompo.gc.ca/Library/54886.pdf 23

Fox, D. S., Amend, M., Merems, A., and Appy, M. (2001). Nearshore rocky reef assessment ROV survey, Newport, OR, Oregon Department of Fish and Wildlife. https://www.dfw.state.or.us/MRP/publications/docs/habitat_2001.pdf 9

Giardino, A., Nederhoff, K., and Vousdoukas, M. (2018). Coastal hazard risk assessment for small islands: Assessing the impact of climate change and disaster reduction measures on Ebeye (Marshall Islands), *Regional Environmental Change*, 18:2237–2248. DOI: 10.1007/s10113-018-1353-3. 61

Giesekam, J., Barrett, J. Taylor, P., and Owen, A. (2014). The greenhouse gas emissions and mitigation options for materials used in UK construction, *Energy and Buildings*, 78:202–2014. DOI: 10.1016/j.enbuild.2014.04.035. 62

Gillett, N. P., Zwiers, F. W., Weaver, A. J., and Stott, P. A. (2003). Detection of human influence on sea-level pressure, *Nature*, 422(6929):292–294. DOI: 10.1038/nature01487. 56

Glover, D. M., Jenkins, W. J., Doney, S. C. (2011). *Modeling Methods for Marine Science*, Cambridge, UK, Cambridge University Press. DOI: 10.1017/cbo9780511975721. 1

Govers, R. (2009). Choking the Mediterranean to dehydration: The Messinian salinity crisis, *Geology*, 37(2):167–170. DOI: 10.1130/g25141a.1. 64

Hall, J. A., Gill, S., Obeysekera, J., Sweet, W., Knuuti, K., and Marburger, J. (2016). Regional sea level scenarios for coastal risk management: Managing the uncertainty of future sea level change and extreme water levels for Department of Defense Coastal Sites Worldwide. Alexandria, VA, U.S. Department of Defense, Strategic Environmental Research and Development Program. 27

Hanson, J. L., Tracy, B. A., Tolman, H. L., and Scott, R. D. (2009). Pacific hindcast performance of three numerical wave models, *Journal of Atmospheric and Oceanic Technology*, 26:1614–1633. DOI: 10.1175/2009jtecho650.1. 10

Hapke, C. J., Kramer, P. A., Fetherston-Resch, E. H., Baumstark, R. D., Druyor, R., Fredericks, X., and Fitos, E. (2019). Florida coastal mapping program—overview and 2018 workshop report. Open File Report 2019–1017, Reston, VA, U.S. Geological Survey. https://pubs.usgs.gov/of/2019/1017/ofr20191017.pdf DOI: 10.3133/ofr20191017. 50

Harrison, W. (1968). Empirical equation for longshore current velocity, *Journal of Geophysical Research*, 73(22):6929–6936. DOI: 10.1029/jb073i022p06929. 66

Hasselmann, K., Barnett, T. P., Bouws, E., Carlson, H., Cartwright, D. E., Enke, K., Ewing, J. A., Gienapp, H., Hasselmann, D. E., Kruseman, P., Meerburg, A., Müller, P., Olbers, D. J., Richter, K., Sell, W., and Walden, H. (1973). Measurements of wind-wave growth and swell decay during the joint north sea wave project. Hamburg, Germany, Deutsches Hydrographisches Institut. https://repository.tudelft.nl/islandora/object/uuid:f204e188--13b9-49d8-a6dc-4fb7c20562fc?collection=research 44

Hawkes, P. J., Gouldby, B. P., Tawn, J. A., and Owen, M. W. (2002). The joint probability of waves and water levels in coastal engineering, *Journal of Hydraulic Engineering*, 40(3):241–251. DOI: 10.1080/00221680209499940. 29

Helsel, D. R. and Hirsch, R. M. (2002). *Statistical Methods in Water Resources, Techniques of Water Resources Investigations*, Book 4, Chapter A3. Reston, VA, U.S. Geological Survey. 50

Hicks, S. D. (2006). Understanding tides. Silver Spring, MD, NOAA National Ocean Service. 21

Horner-Devine, A. R., Hetland, R. D., and MacDonald, D. G. (2015). Mixing and transport in coastal river plumes, *Annual Review of Fluid Mechanics* 47(1):569-594. DOI: 10.1146/annurev-fluid-010313-141408. 26

Hsu, Y. L., Kaihatu, J. M., Dykes, J. D., and Allard, R. A. (2006). Evaluation of Delft3D performance in nearshore flows, NRL memorandum report, NRL/MR/7320–06-8984, Stennis Space Center, MS, Naval Research Laboratory. 25

Hsu, Y. L., Edwards, K. L, and Allard, R. A. (2010). Producing surf forecasting parameters from Delft3D, NRL memorandum report, NRL/MR/7320–10-9213, Stennis Space Center, MS, Naval Research Laboratory. 25

Hyun, B., Baek, S. H., Shin, K., and Choi, K-H. (2017). Assessment of phytoplankton invasion risks in the ballast water of international ships in different growth conditions, *Aquatic Ecosystem Health and Management*, 27(4):423–434. DOI: 10.1080/14634988.2017.1406273. 20

IHO. (2008). *International Hydrographic Organization, IHO Standards for Hydrographic Surveys*. Special Publication no. 44, 5th ed., February 2008, Monaco, International Hydrographic Bureau. 9

Inman, D. L. and Nordstrom, C. E. (1971). On the tectonic and morphologic classification of coasts, *Journal of Geology*, 79:1–21. DOI: 10.1086/627583. 13

Jadid, R., Montoyaa, V. B., and Gabra, M. A. (2020). Effect of repeated rise and fall of water level on seepage-induced deformation and related stability analysis of Princeville levee, *Engineering Geology*, 266(105458). DOI: 10.1016/j.enggeo.2019.105458. 65

James, N. P., Bone, Y., Collins, L. B., and Kyser, T. K. (2001). Superficial sediments of the great Australian bight: Facies dynamics and oceanography on a vast cool-water carbonate shelf, *Journal of Sedimentary Research*, 71(4):549–567. DOI: 10.1306/102000710549. 58

Janzen, C., McCammon, M., Weingartner, T., Statscewich, H., Winsor, P., Danielson, S., and Heim, R. (2019). Innovative real-time observing capabilities for remote coastal regions, *Frontiers in Marine Science*, 6(176). DOI: 10.3389/fmars.2019.00176. 42

Jawak, S. D., Vadlamani, S. S., and Luis, A. J. (2015). A synoptic review on deriving bathymetry information using remote sensing technologies; Models, methods, and comparisons, *Advances in Remote Sensing*, 4(2):146–162. DOI: 10.4236/ars.2015.42013. 9

Jones, C., Raghukumar, K., Friend, P., Scheu, K., and Nichols, C. R. (2019). Assessement of natural hazard vulnerability and resilience in coastal environments, [Day 1 Presentation]. *SERDP and ESTCP Symposium*, Washington, DC. https://serdp-estcp.org/News-and-Events/Conferences-Workshops/2019-Symposium/2019-Symposium-Archive 27

Joyce, S. (2000). The dead zones: Oxygen-starved coastal waters, *Environmental Health Perspectives*, 108(3):A121-A125. DOI: 10.1289/ehp.108-a120. 20

Klemas, V. (2011). Beach profiling and LIDAR bathymetry: An overview with case studies, *Journal of Coastal Research*, 27(6):1019–1028. DOI: 10.2112/jcoastres-d-11-00017.1. 9

Kolker, C. (2003). The Salty Lagoon, *Texas Parks and Wildlife Magazine*, 61(7):50–59. 62

Kraus, N. C. (2007). Coastal inlets of Texas. *Proc. Coastal Sediments '07*, New Orleans, Lam United States, Reston, VA, ASCE Press. DOI: 10.1061/40926(239)114. 59

Kurapov, A. L., Foley, D., Strub, P. T., Egbert, G. D., and Allen, J. S. (2011). Variational assimilation of satellite observation in a coastal model off Oregon, *Journal of Geophysical Research*, 116(C5). DOI: 10.1029/2010jc006909. 10

Laughton, A. S., Gould, W. J., Tucker, M. J., and Roe, H. (2010). *Of Seas and Ships and Scientists: The Remarkable History of the UK's National Institute of Oceanography*, 1949–1973, Cambridge, UK, The Lutterworth Press. 12

Lebreton, L., Slat, B., Ferrari, F., Sainte-Rose, B., Aitken, J., Marthouse, R., Hajbane, S., Cunsolo, S., Schwarz, A., Levivier, A., Noble, K., Debeljak, P., Maral, H., Schoeneich-Argent, R., Brambini, R, and Reisser, J. (2018). Evidence that the great Pacific garbage patch is rapidly accumulating plastic, *Scientific Reports*, 8(1):1–15. DOI: 10.1038/s41598-018-22939-w. 21

Lecours, V. (2017). On the use of maps and models in conservation and resource management (Warning: Results may vary), *Frontiers in Marine Science*, 4(288). DOI: 10.3389/fmars.2017.00288. 47

Lein, J. K. (2012). *Environmental Sensing: Analytical Techniques for Earth Observation*. New York, Springer. DOI: 10.1007/978-1-4614-0143-8. 1

Leuliette, E. W. and Nerem, R. S. (2016). Contributions of Greenland and Antarctica to global and regional sea level change, *Oceanography*, 29(4):154–159. DOI: 10.5670/oceanog.2016.107. 15

Li, W., Isberg, J., Waters, R., Engström, J., Svensson, O., and Leijon, M. (2016). Statistical analysis of wave climate data using mixed distributions and extreme wave prediction, *Energies*, 9(6), Article 396. DOI: 10.3390/en9060396. 29

Limber, P. W., Barnard, P. L., Vitousek, S., and Erikson, L. H. (2018). A model ensemble for projecting multidecadel coastal cliff retreat during the 21st century, *Journal of Geophysical Research: Earth Surface*, 123(7):1566–1589. DOI: 10.1029/2017jf004401. 59

Liu, Y., Kerkering, H., and Weisberg, R. H. (2015). *Coastal Ocean Observing Systems*, Cambridge, MA, Academic Press. DOI: 10.1016/c2014-0-01713-3. 38

Liu, W., Xie, S-P, Liu, Z., and Zhu, J. (2017). Overlooked possibility of a collapsed Atlantic Meridional overturning circulation in warming climate, *Science Advances*, 3(1), e1601666, 1–7. DOI: 10.1126/sciadv.1601666.

Lobe, H. (2015). Recent advances in biofouling protection for oceanographic instrumentation, *Ocean*, 4, Washington DC, MTS/IEEE. DOI: 10.23919/oceans.2015.7401854. 3

Lobe, H., Haldeman, C., and Glenn, S. M. (2010). ClearSignal coating controls biofouling on the rutger glider crossing, *Sea Technology*, 51(5):31–36. 3

Longuet-Higgins, M. S. (1952). On the statistical distribution of the heights of sea waves, *Journal of Marine Research*, 11(3):245–266. DOI: 10.1002/qj.49708135020. 30

Luettich, R. A., Wright, D. L., Signell, R., Friedrichs, C., Friedrichs, M., Harding, J., Fennel, K., Howlett, E., Graves, S., Smith, E., Crane G., and Baltes, R. (2013). Introduction to special section on the U.S. IOOS coastal and ocean modeling testbed, *Journal of Geophysical Research Oceans*, 118:6319–6328. DOI: 10.1002/2013JC008939. 38

Lumpkin, R., Özgökmen, T., and Centurioni, L. (2017). Advances in the application of surface drifters, *Annual Review of Marine Science*, 9:59–81. DOI: 10.1146/annurev-marine-010816-060641. 8

McAllister, M. L., Draycott, S., Adcock, T. A. A, Taylor, P. H., and van den Bremer, T. S. (2019). Laboratory recreation of the Draupner wave and the role of breaking in crossing seas, *Journal of Fluid Mechanics*, 860:767–786. DOI: 10.1017/jfm.2018.886. 29

Menéndez, P., Losada, I. J., Torres-Ortega, S., Narayan, S., and Beck, M. W. (2020). The global flood protection benefits of mangroves, *Scientific Reports*, 10(4404). DOI: 10.1038/s41598-020-61136-6. 69

Mitchell, M. and Bilkovic, D. M. (2019). Embracing dynamic design for climate-resilience living shorelines, *Journal of Applied Ecology*, 56(5):1099–1105. DOI: 10.1111/1365-2664.13371. 66

Morrow, R. and Le Traon, P.-Y. (2012). Recent advances in observing mesoscale ocean dynamics with satellite altimetry, *Advances in Space Research*, 50(8):1062–1076. DOI: 10.1016/j.asr.2011.09.033. 15

Mueller, D., Uibel, S., Takemura, M., Klingelhoefer, D., and Groneberg, D. A. (2011). Ships, ports, and particulate air pollution—an analysis of recent studies, *Journal of Occupational Medicine and Toxicology*, 6(31). DOI: 10.1186/1745-6673-6-31. 56

National Research Council. (2003). *Government Data Centers: Meeting Increasing Demands*. Washington, DC, National Academies Press. 5

Neumann, G. and Pierson, W. J. (1957). A detailed comparison of theoretical wave spectra and wave forecasting methods, *Deutsche Hydrographische Zeitschrift*, 10:134–146. DOI: 10.1007/bf02019597. 44

Nichols, C. R. (1993). Operational characteristics of the Tampa Bay physical oceanographic real-time system, in S. S. Y. Wang, (Ed.), *Advances in HydroScience and Engineering*, 1(part B):1491–1498. University, MS, Center for Computational Hydroscience and Engineering, School of Engineering, University of Mississippi. 7, 22, 42

Nichols, C. R. (2003). Coastal awareness and preparedness, *Sea Technology Magazine*, 44(2):85, Arlington, VA, Compass Publications, Inc. 12

Nichols, C. R., Luettich, R. A., Wright, L. D., Friedrichs, M. A. M., Kurapov, A., van der Westhuysen, A. J., Fennel, K., Howlett, E., Akli, A., and Crane, G. (2017). Coastal and ocean modeling testbed data management plan, Washington, DC, Southeastern Universities Research Association. https://ioos.us/images/comt-dm-plan.pdf 38

Nichols, C. R. and L. J. Pietrafesa (1997). Oregon Inlet: Hydrodynamics, volumetric flux and implications for larval fish transport, *Report No. DOE/ER/61425-T3*, Washington, DC, US-DOE Office of Scientific and Technical Information, DOI: 10.2172/479074. 41

Nichols, C. R. and Williams, R. G. (2017). *Encyclopedia of Marine Science*, 2nd ed., New York, Infobase Publishing. 55

Nichols, C. R., Wright, L. D., Bainbridge, S. J., Cosby, A. Hénaff, A., Loftis, J. D., Cocquempot, L., Katragadda, S., Mendez, G. R., Letortu, P., Le Dantec, N., Resio, D., and Zarillo, G. (2019). Collaborative science to enhance coastal resilience and adaptation, *Frontiers in Marine Science*, 6(404). DOI: 10.3389/fmars.2019.00404. 12

NOAA. (2005). *2nd Workshop Report on the Quality Assurance of Real-Time Ocean Data*, Norfolk, VA, NOAA/NOS/Center for Operational Oceanographic Products and Services, 48pp, (CCPO Technical Report Series no. 05–01).

Noufal, K. K., Najeem, S., and Venkatesan, L. R. (2017). Seasonal and long term evolution of oceanographic conditions based on year-around observation in Kongsfjorden, Arctic, *Polar Science*, 11:1–10. DOI: 10.1016/j.polar.2016.11.001. 41

Ochi, M. K. and Hubble, E. N. (1976). Six-parameter wave spectra. *Proc. of the 15th Coastal Engineering Conference*, pages 301–328, Honolulu, HI, NY, American Society of Civil Engineers. DOI: 10.9753/icce.v15.17. 44

Open University. (2001). *Ocean Circulation*, 2nd ed., Boston, MA, Butterworth-Heinemann. 71

Parker, B. B. (2007). Tidal analysis and prediction. *NOAA Special Publication NOS CO-OPS 3*, Silver Spring, MD, NOAA NOS Center for Operational Oceanographic Products and Services. 21

Pasman, I., Kurapov, A. L., Barth, J. A., Kosro, P. M., and Shearman, R. K. (2019). Why gliders appreciate good company: Glider assimilation in the Oregon–Washington coastal ocean 4DVAR system with and without surface observations, *Journal Geophysical Research Oceans*, 124(1):750–772. DOI: 10.1029/2018jc014230. 10

Pawlowicz, R., Beardsley, B., and Lentz, S. (2002). Classical tidal harmonic analysis including error estimates in MATLAB using T_TIDE, *Computers and Geosciences*, 28(8):929–937. DOI: 10.1016/s0098-3004(02)00013-4. 41

Pickard, G. L. and Emery, W. J. (1982). *Descriptive Physical Oceanography: An Introduction*, 4th ed., Oxford, UK, Pergamon Press. 20

Pietrafesa, L. J., Blanton, J. O., Wang, J. D., Kourafalou, V. H., Lee, T. N., and Bush, K. A. (1985). The tidal regime in the South Atlantic Bight, in Atkinson, L. P., Menzel, D. W., and Bush, K. A. (Eds.), *Oceanography of the Southeastern U.S. Continental Shelf*. pp. 63–76, Washington, DC, American Geophysical Union. 24

Pietrafesa, L. J. and Janowitz, G. S. (1988). Physical oceanographic processes affecting larval transport around and through North Carolina inlets, *American Fisheries Society Symposium*, 3:34–50. 55

Pilkey, O. H., Neal, W. J., Riggs, S. R., Webb, C. A., Bush, D. M., Pilkey, D. F., Bullock, J., and Cowan, B. A. (1998). *The North Carolina Shore and its Barrier Islands: Restless Ribbons of Sand*. Durham, NC, Duke University Press. 57

Portilla, J., Ocampo-Torres, F. J., and Monbaliu, J. (2009). Spectral partitioning and identification of wind sea and swell, *Journal of Atmospheric and Oceanic Technology*, 26(1):107–122. DOI: 10.1175/2008jtecho609.1. 44

Pugh, D. T. (1987). *Tides, Surges and Mean Sea Level: A Handbook for Engineers and Scientists*. Chichester, West Sussex, UK, John Wiley & Sons. 1

Raghukumar, K., Chang, G., Spada, F., Jones, C., Janssen, T., and Gans, A. (2019). Performance characteristics of spotter, a newly developed real-time wave measurement buoy, *Journal of Atmospheric and Oceanic Technology*, 36(6):1127–1141. DOI: 10.1175/jtech-d-18-0151.1. 44, 45

Raineault, N. A., Gee, L., Kane, R., Saunders, M., Heffron, E., and Carey, S. (2018). E/V Nautilus mapping summary vol. 2017: Cascadia margin to the Revillagigedo Archipelago and beyond, *Oceanography*, 31(1), Supplement, 26–27. 9

Raineault, N. A. and Ballard, R. D. (2018). Nautilus field season overview, *Oceanography*, 31(1), Supplement, 24–25. 9

Rao, Y. R., Hawley, N., Charlton, M. N., and Schertzer, W. M. (2008). Physical processes and hypoxia in the central basin of Lake Erie, *Limnology and Oceanography*, 53(5):2007–2020. DOI: 10.4319/lo.2008.53.5.2007. 63

Reeve, D., Chadwick, A., and Fleming, C. (2018). *Coastal Engineering: Processes, Theory and Design Practice*, 3rd ed., Boca Raton, FL, CRC Press. DOI: 10.1201/b11804. 12

Rogerson, S. (2019). Transformation of the argos data collection system in the 2020s [poster presentation]. *NOAA Emerging Technologies Workshop*, College Park, MD. https://nosc.noaa.gov/public_docs/NOAA_2019_ETW_report.pdf 7

Sampson, C. R., Hansen, J. A., Wittmann, P. A., Knaff, J. A., and Schumacher, A. (2016). Wave probabilities consistent with official tropical cyclone forecasts, *Weather and Forecasting*, 35(2):2035–2045. DOI: 10.1175/waf-d-15-0093.1. 64

Schnyder, J. S. D., Eberli, G. P., Kirby, J. T., Shi, F., Tehranirad, B., Mulder, T., Ducassou, E., Hebbeln, D., and Wintersteller, P. (2016). Tsunamis caused by submarine slope failures along western Great Bahama Bank, *Scientific Reports*, 6(35925). DOI: 10.1038/srep35925. 70

Shepard, F. P. (1937). Revised classification of marine shorelines, *Journal of Geology*, 45(6):602–624. DOI: 10.1086/624584. 13

Shepard, F. P. (1973). *Submarine Geology*. New York, Harper and Row. 13

Shope, J. B., Storlazzi, Erikson, L. H., and Hagermiller, C. A. (2016). Changes to extreme wave climates of islands within the Western Tropical Pacific throughout the 21st century under RCP 4.5 and RCP 8.5, with implications for island vulnerability and sustainability, *Global and Planetary Change*, 141:25–38. DOI: 10.1016/j.gloplacha.2016.03.009. 30

Showstack, R. (2014). Scientists focus on land subsidence impacts on coastal and delta cities, *Eos, Transactions American Geophysical Union*, 95(19):59. DOI: 10.1002/2014eo190003. 65

Shureman, P. (1958). *Manual of Harmonic Analysis and Prediction of Tides*. Special Publication no. 98, U.S. Coast and Geodetic Survey, Washington, DC, U.S. Government Printing Office. DOI: 10.5962/bhl.title.38116. 21

Smith, W. H. F. and Sandwell, D. T. (2004). Conventional bathymetry, bathymetry from space, and geodetic altimetry, *Oceanography*, 17(1):8–23. DOI: 10.5670/oceanog.2004.63. 9

Song, H., Edwards, C. A., Moore, A. M., and Fiechter, J. (2016a). Data assimilation in a coupled physical-biogeochemical model of the California current system using an incremental lognormal 4-dimensional variational approach: Part 1—model formulation and biological date data assimilation twin experiments, *Ocean Modelling*, 106:131–145. DOI: 10.1016/j.ocemod.2016.04.001. 10

Song, H., Edwards, C. A., Moore, A. M., and Fiechter, J. (2016b). Data assimilation in a coupled physical-biogeochemical model of the California current system using an incremental lognormal 4-dimensional variational approach: Part 2—joint physical and biological data assimilation twin experiments, *Ocean Modelling*, 106:146–158. DOI: 10.1016/j.ocemod.2016.09.003. 10

Sowers, D. C., Masetti, G., Mayer, L. A., Johnson, P., Gardner, J. V., and Armstrong, A. A. (2020). Standardized geomorphic classification of seafloor within the United States Atlantic canyons and continental margin, *Frontiers in Marine Science*, 7(9). DOI: 10.3389/fmars.2020.00009.

Sweet, W. V., Park, J. C., Marra, J. J., Zervas, C., and Gill, S. (2014). Sea level rise and nuisance flood frequency changes around the United States. *NOAA Technical Report NOS CO-OPS 073*. National Oceanic and Atmospheric Administration, National Ocean Service, Silver Spring, MD. DOI: 10.13140/2.1.3900.2887. 68

Talke, S. A. and Jay, D. A. (2013). 19th century North American and Pacific tidal data: Lost or just forgotten, *Journal of Coastal Research*, 29(6A):118–127. DOI: 10.2112/jcoastres-d-12-00181.1. 21

Toon, J. (November 23, 2016). Flying High: GTRI acquires TigerShark unmanned aerial vehicles, *Georgia Tech Research Horizons, 3*, https://rh.gatech.edu/front-office/flying-high 7

Tylkowski, J. and Hojan, M. (2019). Time decomposition and short-term forecasting of hydrometeorological conditions in the south Baltic coastal zone of Poland, *Geosciences*, 9(2), Article 68. DOI: 10.3390/geosciences9020068. 41

U.S. Army Corps of Engineers. (2002). Coastal engineering manual (CEM), *Engineer Manual 1110–2-1100*, Washington, DC, U.S. Army Corps of Engineers, (6 volumes). 55

U.S. Commission on Ocean Policy. (2004). *An Ocean Blueprint for the 21st Century: Final Report*. Washington, DC, U.S. Commission on Ocean Policy. 18

U.S. Army Corps of Engineers. (2013). *Incorporating Sea-Level Change in Civil Works Programs*, Regulation no. 1100–2-8162, Washington, DC, Department of the Army. http://www.publications.usace.army.mil/Portals/76/Publications/EngineerRegulations/ER_1100-2-8162.pdf 27

Viana, M., Hammingh, P., Colette, A., Querol, X., Degraeuwe, B., Vlieger, I., and Aerdenne, J. (2014). Impact of maritime transport emissions on coastal air quality in Europe, *Atmospheric Environment*, 90:96–105. DOI: 10.1016/j.atmosenv.2014.03.046. 20

Voermans, J. J., Smit, P. B., Janssen, T. T., and Babanin, A. V. (2020). Estimating wind speed and direction using wave spectra, *Journal of Geophysical Research Oceans*, 125(2), Article e2019JC015717. DOI: 10.1029/2019jc015717. 42

Waaijers, L. and van der Graaf, M. (2011). Quality of research, an operational approach, *D-Lib Magazine*, 17(1/2). DOI: 10.1045/january2011-waaijers. 39

Wahl, T., Jain, S., Bender, J., Meyers, S. D., and Luther, M. E. (2015). Increasing risk of compound flooding from storm surge and rainfall for major U.S. cities, *Nature Climate Change*, 5(12):1093–1097. DOI: 10.1038/nclimate2736. 60

Walters, D. C. and Kirwin, M. L. (2016). Optimal hurricane overwash thickness for maximizing marsh resilience to sea level rise, *Ecology and Evolution*, 6(9):2948–2956. DOI: 10.1002/ece3.2024. 58

Wang, C., Corbett, J. J., and Firestone, J. (2008). Improving spatial representation of global ship emissions inventories, *Environmental Science Technology*, 42(1):193–199. DOI: 10.1021/es0700799. 3

Ward, P. J., Jongman, B., Kummu, M., Dettinger, M. D., Weiland, F. C. S., and Winsemius, H. C. (2014). Strong influence of el niño southern oscillation on flood risk around the world, *Proc. of the National Academy of Sciences of the United States of America*, 111(44):15659–15664. DOI: 10.1073/pnas.1409822111. 61

Weber, J. E. (1983). Steady wind-and wave-induced currents in the open ocean, *Journal of Physical Oceanography*, 13(3):524–530. DOI: 10.1175/1520-0485(1983)013%3C0524:swawic%3E2.0.co;2. 71

Whitt, A. D., Dudzinski, K., and Laliberté, J. R. (2013). North Atlantic right whale distribution and seasonal occurrence in nearshore waters off New Jersey, USA, and implications for management, *Endangered Species Research*, 20:59–69. DOI: 10.3354/esr00486. 25

Williams, R. G., French, G. W., and Nichols, C. R. (1993). Nowcasting of currents in Tampa Bay using a physical oceanographic real-time system, in S. S. Y. Wang, (Ed.), *Advances in HydroScience and Engineering*, 1(part B), 1507–1512, University, MS, Center for Computational Hydroscience and Engineering, School of Engineering, University of Mississippi. 42

Williams, J. J. and Esteves, L. S. (2017). Guidance on setup, calibration, and validation of hydrodynamic, wave, and sediment models for shelf seas and estuaries. *Advances in Civil Engineering*, Article 5251902. DOI: 10.1155/2017/5251902. 30

Wolanski, E. (2014). *Estuaries of Australia in 2050 and Beyond*. Dordrecht, Netherlands, Springer. DOI: 10.1007/978-94-007-7019-5. 62

World Meteorological Organization. (2018). *Guide to Marine Meteorological Services*, 2018 ed., WMO-No. 471. Geneva, Switzerland, WMO. https://library.wmo.int/doc_num.php?explnum_id=5445 3

Wright, L. D., Nichols, C. R., Cosby, A. G., and D'Elia, C. F. (2016). Collaboration to enhance coastal resilience, *Eos, Transactions American Geophysical Union*, 97. http://bit.ly/2bwBwrU DOI: 10.1029/2016eo057981. 27

Wright, L. D. and Nichols, C. R. (2019). *Tomorrow's Coasts: Complex and Impermanent. Coastal Research Library Series*, 27, Cham, Switzerland, Springer. DOI: 10.1007/978-3-319-75453-6. 12, 27, 55, 56

Xia, M., Mao, M., and Niu, Q. (2020). Implementation and comparison of the recent three-dimensional radiation stress theory and vortex-force formalism in an unstructured-grid coastal circulation model, *Estuarine, Coastal and Shelf Science,* 240(106771). DOI: 10.1016/j.ecss.2020.106771. 25

Xiao, H.-W, Xiao, H.-Y, Shen, C.-Y, Zhang, Z.-Y, and Long, A.-M. (2018). Chemical composition and sources of marine aerosol over the western north pacific ocean in winter, *Atmosphere*, 9(8), Article 298. DOI: 10.3390/atmos9080298. 55

Xiong, C., Li, Z., Sun, X., Zhai, J., and Niu, Y. (2018). An effective method for submarine pipeline inspection using Three-Dimensional (3D) models constructed from multisensor data fusion, *Journal of Coastal Research*, 34(4):1009–1019. DOI: 10.2112/jcoastres-d-17-00109.1. 42

Zalasiewicz, J., Williams, M., Smith, A., Barry, T. L., Coe, A. L., Bown, P. R., Brenchley, P., Cantrill, D., Gale, A., Gibbard, P., Gregory, F. J., Hounslow, M. W., Kerr, A. C., Pearson, P., Knox, R., Powell, J., Waters, C., Marshall, J., Oates, M., Rawson, P., and Stone, P. (2008). Are we now living in the Anthropocene?, *GSA Today*, 18(2):4–8. DOI: 10.1130/gsat01802a.1.

Zervas, C. (1999). Tidal current analysis procedures and associated computer programs. *NOAA Technical Memorandum NOS CO-OPS 0021*. Silver Spring, MD, Center for Operational Oceanographic Products and Services Products and Services Division, NOAA. https://tidesandcurrents.noaa.gov/publications/techrpt21.pdf 21

Zetler, B. D. (1982). Computer applications to the tides in the national ocean survey: Supplement to manual of harmonic analysis and prediction of tides, *NOAA Special Publication*, 98, Silver Spring, MD, NOAA. 21

Zhang, A., Hess, K. W., and Aikman III, F., (2010). User-based skill assessment techniques for operational hydrodynamic forecast systems, *Journal of Operational Oceanography*, 3(2):11–24. DOI: 10.1080/1755876x.2010.11020114. 31

Zhao, D, Cheng, X, Zhang, H., Niu, Y., Qi, Y., and Zhang, H. (2018). Evaluation of the ability of spectral indices of hydrocarbons and seawater for identifying oil slicks utilizing hyperspectral images, *Remote Sensing*, 10(3):421. DOI: 10.3390/rs10030421. 7

Authors' Biographies

C. REID NICHOLS

C. Reid Nichols is the president of Marine Information Resources Corporation. He is a retired U.S. Marine Corps colonel with experience in military logistics and intelligence. His applied research interests include physical oceanography and providing environmental characterization to support researchers such as naval architects involved in testing innovative vessels, engineers involved in the development of oceanographic platforms, and operators interested in decision aid products that help protect property and save lives. He received his M.S. from North Carolina State University and an MBA from Loyola University Maryland and was a contributing author for *Recent Advances and Issues in Oceanography*, *Encyclopedia of Marine Science*, and *Tomorrow's Coasts: Complex and Impermanent*.

KAUSTUBHA RAGHUKUMAR

Kaustubha Raghukumar is a physical oceanographer with Integral Consulting, Inc. His research has focused on modeling ocean ecosystems such as the California Current System using data-constrained physical circulation estimates and validation testing of innovative instruments such as the *Spotter*, an easily deployed solar powered wave buoy. He received his M.S. from the New Jersey Institute of Technology and a Ph.D. from University of California San Diego.

Printed in the United States
by Baker & Taylor Publisher Services